Anatomy of Movement

ANATOMY

—— OF ——

Movement

REVISED EDITION

Text and illustrations by

Blandine Calais-Germain

EASTLAND PRESS ❖ SEATTLE

Originally published as *Anatomie pour le mouvement,* Editions Desiris (France), 1985. Revised in 1991 and 1999

English language edition © 1993, 2007 by Eastland Press, Inc.
P.O. Box 99749
Seattle, WA 98139, USA
www.eastlandpress.com

Publisher's Cataloging-In-Publication Data

Calais-Germain, Blandine.
    [Anatomie pour le mouvement. English]
    Anatomy of movement / text and illustrations by Blandine Calais-Germain ; edited by John O'Connor and Allan Kaplan. — Rev. ed. / translated by Regine MacKenzie.

        p. : ill. ; cm.

    Includes bibliographical references and index.
    Originally published in French: [Paris] : Editions DesIris, 1985. Rev. 1991 and 1999, under title: Anatomie pour le mouvement.

    ISBN: 978-0-939616-57-2

    1. Musculoskeletal system. 2. Human mechanics. 3. Kinesiology. I. O'Connor, John. II. Kaplan, Allan. III. MacKenzie, Regine. IV. Title. V. Title: Anatomie pour le movement.

QP301 .C2313 2007
612.7/6

                                            2006936656

        6 8 10 9 7

Translated by Regine MacKenzie
Edited by John O'Connor and Allan Kaplan
Cover illustration by Sandy Johnson
Cover design by Patricia O'Connor

Book design by Gary Niemeier

Dedicated to Marie, Patrick, Jacques, Francois

Table of Contents

··

1

Foreword

Anatomists, for many centuries, were concerned almost exclusively with precise description of the body's structures. Inevitably, they began by treating the locomotor system in the same way as the internal organs, i.e., actual functions were either unknown or described independently of structure.

Gradually, around the beginning of the 20th century, anatomists began paying more attention to the actions of muscles and joints. Such functional studies remained at an elementary level for several decades. More recently, some researchers began looking at biomechanical properties (such as elasticity and resistance) of the locomotor system. However, these studies were focused on isolated components in the laboratory, not on how muscles and joints are used in "real life." Functional aspects were often viewed in terms of "efficiency," i.e., how to make the body an obedient instrument of various physical disciplines.

In physiotherapy, body movements are analyzed in terms of both neurophysiological and mechanical components, thus allowing better definition of therapeutic effects and the real mechanisms of movement.

Many people interested in physical disciplines such as dance, mime, theater, yoga, relaxation, etc. have come to physiotherapy looking for quantitative as well as qualitative analytical studies which would facilitate their practice. In this way, Blandine Calais-Germain began by studying dance and ended up studying physiotherapy.

The complementary nature of these two ways of dealing with human body movements is obvious. Blandine quickly realized that dancers could benefit greatly from a better understanding of their "inner" bodies. She devised a novel teaching method to serve this purpose: the simultaneous representation of physical structures and their possible movements, designed to facilitate actual execution by the dancer.

Not only dancers, but also professionals involved in other physical disciplines, came in increasing numbers to her classes. The emphasis in these classes (and this book) is on anatomy not for its own sake, but for better understanding of body movements.

I have taken great pleasure in witnessing the birth of this concept, the first classes, and now the publication of this book which embodies Blandine's many years of experience as a dancer and teacher. I am delighted that the fruits of this experience will now be made widely available to others. Having worked closely with Blandine when she was a student of physiotherapy, I can attest to her skills as a therapist, her intelligence, and her love for teaching.

The drawings in this book are all original, and the emphasis is always on description and understanding of natural postures and movements. The book will be particularly useful to those who, by profession, deal with integrated or complex movements of the body. For those who deal with human anatomy in any way, it will provide a useful and thought-provoking resource. I wish for this book the great success it deserves.

Dr. Jacques Samuel
Director, French School of Orthopedics and Massage
Paris, France

Preface

. . . .

I would like to briefly describe the content and organization of this book.

This is intended simply as an introductory text. The emphasis is on basic human anatomy as it relates to *external body movement.* Therefore, we will be concerned mainly with bones, muscles, and joints. There will be no description of the skull, visceral organs, circulatory system, central nervous system, etc.

The book is designed to be as compact as possible, and to avoid repetition. Thus, format may vary from one chapter to the next. Parts of the body that are affected by the same muscles may be described together. Reference may be given to a previous page where a certain structure or function is described in more detail.

For consistency and ease of orientation, drawings usually show structures from the right side of the body. Exceptions are clearly indicated.

Joints are sometimes drawn without the adjacent bones, so that the articular surfaces can be more clearly seen. Similarly, each muscle is drawn in isolation (without surrounding muscles) to make its function more obvious.

Chapter 1 provides basic orientation and terminology, and should be read first. Subsequent chapters are arranged in a logical order (starting with the trunk, moving down the arm, and then down the leg), and I recommend that they be read in this order. However, the reader with previous knowledge of anatomy may start at any chapter.

The index will be helpful for locating the page where a particular structure is first mentioned, or described in detail.

CHAPTER ONE

Introduction

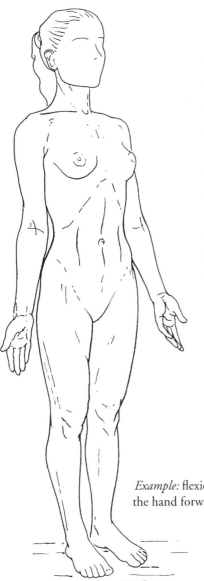

Anatomical position

The anatomy of movement, in the human body, mainly involves interaction of three systems:

- the **bones** of the skeleton,
- linked together at the **joints,**
- are moved by action of the **muscles.**

Description of movements can be difficult. Various parts of the body can move in many different directions. Often more than one joint is involved. For consistency, the following conventions are generally followed:

- we begin by considering each joint in isolation;
- three perpendicular planes are used for reference;
- movements are described in relation to a standard "anatomical position" in which the body is standing upright, the feet parallel, the arms hanging by the sides, and the palms and face directed forward (see illustration). This position is not a common posture, but it is helpful for reference purposes to describe the starting point of a movement.

Example: flexion of the wrist is a movement that takes the hand forward from the anatomical position

Planes of movement

The **median plane** divides the body into symmetrical right and left halves.

Any plane parallel to the median plane is called a **sagittal plane**. Movements in this plane can be seen from the side.

A movement in a sagittal plane which takes a part of the body forward from anatomical position is called **flexion.**

Example: flexion of the hip

Example: flexion of the shoulder

Exception: flexion (dorsiflexion) of the ankle

A movement in a sagittal plane which takes a part of the body backward from anatomical position is called **extension.**

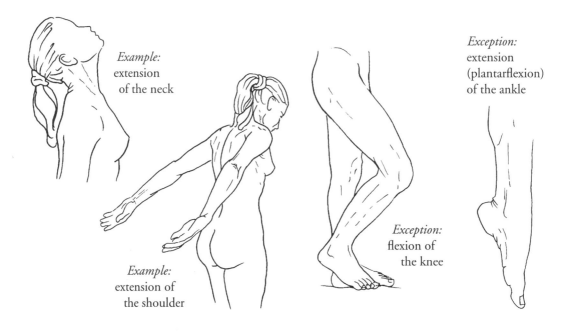

Example: extension of the neck

Exception: extension (plantarflexion) of the ankle

Example: extension of the shoulder

Exception: flexion of the knee

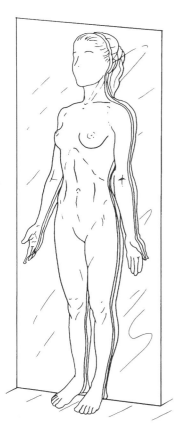

A coronal or **frontal plane** is any plane perpendicular to the median plane. It divides the body into anterior and posterior parts.

A movement in a frontal plane which takes a part of the body toward the median plane is called **adduction.**

Movements on the frontal plane can be seen from the front.

The opposite type of movement (away from the median plane) is called **abduction.**

Example: adduction of the hip

Example: abduction of the shoulder

For the trunk or neck, movement in the frontal plane away from the median plane is called **lateral flexion** or **sidebending.**

Example: right lateral flexion of the trunk

For the fingers or toes, the reference used is the axis of the hand (middle finger) or foot (2d toe).

Example: abduction of the fifth finger moves it away from the axis of the hand (not from the median plane)

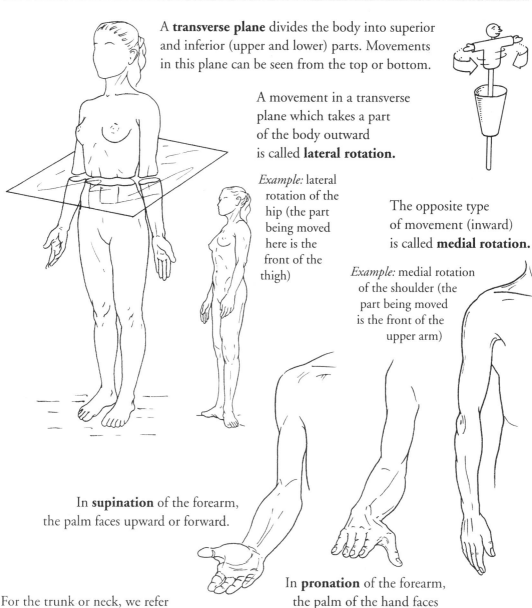

A **transverse plane** divides the body into superior and inferior (upper and lower) parts. Movements in this plane can be seen from the top or bottom.

A movement in a transverse plane which takes a part of the body outward is called **lateral rotation.**

Example: lateral rotation of the hip (the part being moved here is the front of the thigh)

The opposite type of movement (inward) is called **medial rotation.**

Example: medial rotation of the shoulder (the part being moved is the front of the upper arm)

In **supination** of the forearm, the palm faces upward or forward.

In **pronation** of the forearm, the palm of the hand faces downward or backward.

For the trunk or neck, we refer simply to right or left rotation. The reference point is the front of the chest or head.

Complex body movements typically involve movement in all three planes.

Example: sitting in the "tailor's position" involves flexion, abduction, and lateral rotation of the hip joints

Other anatomical reference terms

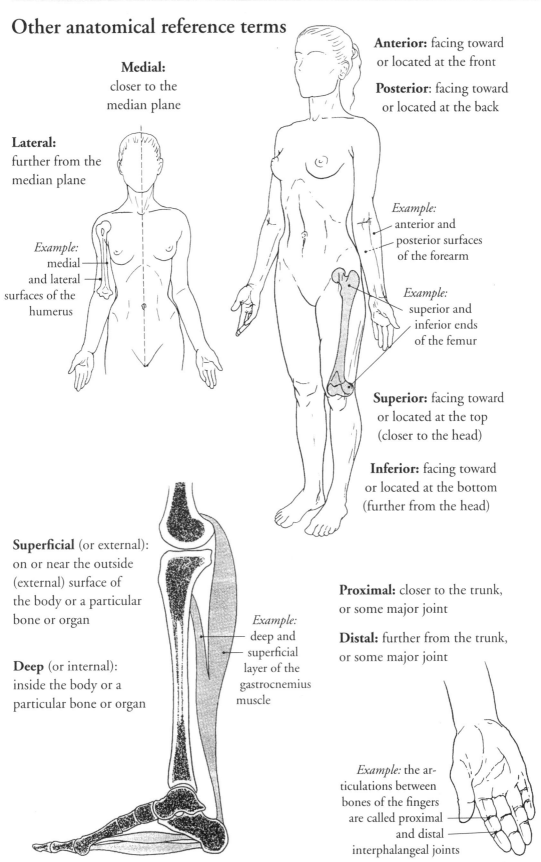

Medial: closer to the median plane

Lateral: further from the median plane

Example: medial and lateral surfaces of the humerus

Anterior: facing toward or located at the front

Posterior: facing toward or located at the back

Example: anterior and posterior surfaces of the forearm

Example: superior and inferior ends of the femur

Superior: facing toward or located at the top (closer to the head)

Inferior: facing toward or located at the bottom (further from the head)

Superficial (or external): on or near the outside (external) surface of the body or a particular bone or organ

Deep (or internal): inside the body or a particular bone or organ

Example: deep and superficial layer of the gastrocnemius muscle

Proximal: closer to the trunk, or some major joint

Distal: further from the trunk, or some major joint

Example: the articulations between bones of the fingers are called proximal and distal interphalangeal joints

Skeleton

The skeleton is a mobile framework of bones providing rigid support for the body. The bones also serve as levers for the action of muscles.

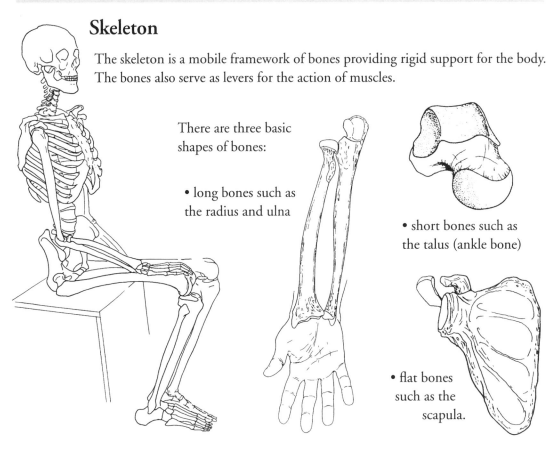

There are three basic shapes of bones:

• long bones such as the radius and ulna

• short bones such as the talus (ankle bone)

• flat bones such as the scapula.

Bone tissue consists of about two-thirds mineral components (mostly calcium salts), which give rigidity, and one-third organic components, which give elasticity. Both these qualities are essential. Without rigidity bones would not keep their shape, but without some elasticity they would shatter too easily.

Bones are subjected to several types of mechanical strain:

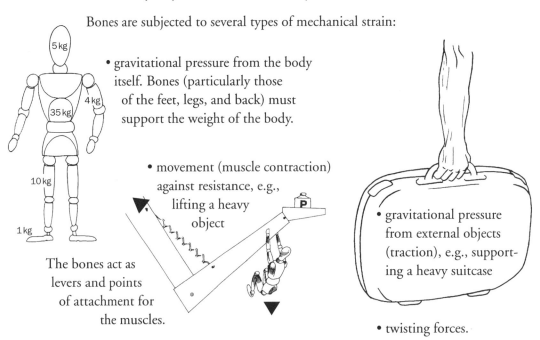

• gravitational pressure from the body itself. Bones (particularly those of the feet, legs, and back) must support the weight of the body.

5 kg
4 kg
35 kg
10 kg
1 kg

• movement (muscle contraction) against resistance, e.g., lifting a heavy object

P

The bones act as levers and points of attachment for the muscles.

• gravitational pressure from external objects (traction), e.g., supporting a heavy suitcase

• twisting forces.

Bones

The structure of the bones shows us that
they are made to withstand all these types of strain.
We can see how by looking at a bone in cross section.

A long bone,
in this case the femur,
consists of three parts.

• The central shaft is
called the **diaphysis.**

• The two ends are
called the **epiphyses.**

*trabecular
structure*

*bone
marrow*

diaphysis

periosteum

The diaphysis is a hollow tube with walls made of compact
bone. The hollow structure gives light weight and is actually
sturdier than a solid structure would be. Compact bone is
thickest in the middle section of the diaphysis where mechanical
strains are greatest. It also predominates in the concave areas
of the curved ends.

A cross section of the epiphysis shows a **trabecular** (spongy)
structure. Fibers are arranged in rows along the lines
of greatest mechanical stress.

The hollow part of the diaphysis contains the **bone marrow**.
The marrow is red in children, but becomes yellow in adults
as much of it is replaced by fatty tissue.

The external surface of the bone is covered with a membrane
called the **periosteum**.

The articular surfaces of the bones are covered with
articular (hyaline) cartilage.

articular cartilage

Joints

Joints are areas where bones are linked together. They have varying degrees of mobility.

In some joints the bones are linked simply by fibrous connective tissue or cartilage. This allows little or no movement. These joints are not of great interest in a book about movement, but we will mention them occasionally.

Our primary interest will be in freely-movable joints called **diarthroses**, or **synovial joints**. These surfaces (sometimes called facets) are shaped so as to fit together but also allow movement. There are many general categories of joints, based on the shape of the articulating surfaces.

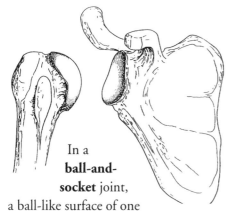

In a **ball-and-socket** joint, a ball-like surface of one bone fits into a cup-like depression of another bone. This type of joint, also called an enarthrodial or spheroidal joint, allows movement in all directions.

Examples: shoulder

Ellipsoid: Similar to the joint on the left, the shallow cavity of one bone receives the rounded surface of another. This allows movement in all three planes described on pages 8-10.

Examples: metacarpophalangeal

Hinge: The convex surface of one bone fits against the concave surface of another. Movement is chiefly in one plane.

Examples: ankle

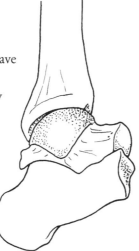

Pivot: The rounded surface of one bone fits into a ring formed partly by another bone. Movement is chiefly in one direction, like the hinge of a door.

Examples: radioulnar

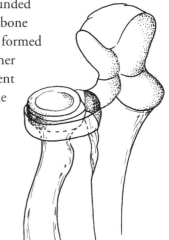

Saddle: Both surfaces are saddle-shaped, i.e., convex in one direction and concave in the other.

This permits movement in two planes.

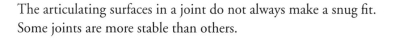

Example: sterno-clavicular

The articulating surfaces in a joint do not always make a snug fit. Some joints are more stable than others.

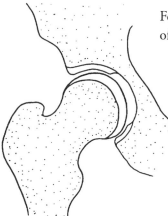

For example, the ball-and-socket structure of the hip is deep and snug-fitting.

By contrast, the ball-and-socket of the shoulder is shallow, looser and less stable.

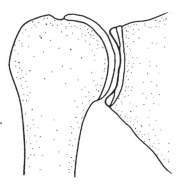

Between the articulating ends of the two bones in a joint is a gap which is not opaque to X-rays. This corresponds to the articular cartilages and synovial cavity. When you look at the X-ray, you can only see the thickness of the cartilages in the joint. The cartilages themselves are not visible on X-ray. Thus, it looks like there is a free space between the two bones.

In a dislocation or subluxation, a bone is moved from its normal position in a joint because of some trauma. There is associated damage to ligaments, etc.

Cartilage

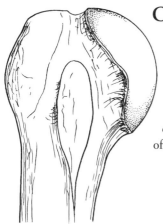

Articulating surfaces of bones are covered with a shiny, whitish connective tissue called cartilage. It contributes to the synovial capsule and also protects the underlying bone tissue.

Example: cartilage of the head of the humerus

When movement occurs, joint cartilage is subjected to two types of stress:

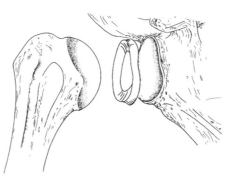

gravitational pressure (particularly in the weight-bearing joints of the legs and feet),

Cartilage is well-adapted to these stresses, being strong, resilient, and smooth. Thus it can allow some sliding of the bones relative to each other.

5 kg

35 kg   4 kg

10 kg

1 kg

…and friction from the movement itself.

Nevertheless, cartilage may be damaged either by trauma or excessive wear (e.g., when the ends of the articulating bones do not provide a good "fit"). Rheumatoid arthritis and osteoarthritis are two common diseases involving damage to joint cartilage, accompanied by inflammation, pain, and stiffness of the joint and surrounding muscles.

Joint cartilage (like all cartilage) does not contain blood vessels. It receives nutrients from the synovial fluid and the bone it covers.

Fibrocartilage contains high concentrations of collagenous (white) fibers and is specially adapted for absorbing shock. It is found in the intervertebral discs and in the menisci (articular discs) of the knee and other large joints.

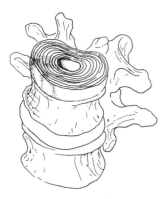

intervertebral discs

fibrocartilagenous padding
(e.g., in shoulder joint)

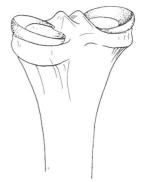

menisci

Joint capsule

This sleeve-like structure encloses the joint, prevents loss of fluid, and binds together the ends of the articulating bones. The outer layer of the capsule is composed of dense connective tissue and represents a continuation of the periosteum.

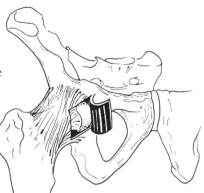

Example: in the hip joint we have opened a "window" into the capsule

The joint capsule is stronger where movement must be prevented.

Example: the knee joint (starting from anatomical position) allows only flexion. The capsule is strongly reinforced posteriorly to prevent extension.

Example: anterior ligaments of the hip joint

Fibers of the outer capsule are often arranged in parallel bundles (called **ligaments;** see following section) to reinforce joints and prevent unwanted movement.

The capsule may be arranged loosely or in folds where movement is possible.

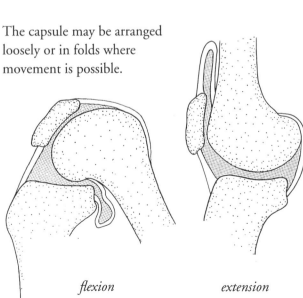

flexion   *extension*

Example: the capsule of the knee is loose in the front to allow flexion, and becomes folded during extension

The interior of the capsule is covered by a **synovial membrane**, which covers the deep surface of the capsule and folds over at the capsular insertions. Its principal function is to secrete **synovial fluid** (shown as gray in the drawing) that fills the articular cavity.

This fluid lubricates the joint and provides nutrients to the cartilage.

Ligaments

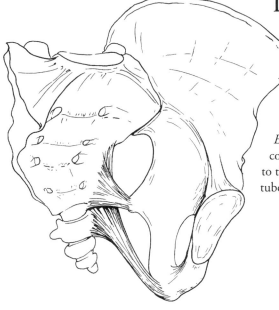

Ligaments are dense bundles
of parallel collagenous fibers.
They are often derived from the outer
layer of the joint capsule, but may also
connect nearby but non-articulating bones.

Example: the ligament
connecting the sacrum
to the spine and
tuberosity of the ischium

The ligaments function chiefly to strengthen and stabilize the
joint in a passive way. Unlike the muscles, they cannot actively
contract. Nor (except for a few ligaments which contain a high
proportion of yellow elastic fibers) can they stretch.

extension   *flexion*

They are placed under tension
by certain positions of the
joint and slackened by
others.

Example: the fibular (lateral) collateral ligament of the
knee is pulled tight in extension, and loosened in flexion

Ligaments contain numerous sensory
nerve cells capable of responding to
the speed, movement, and
position of the joint, as well
as to stretching or pain.

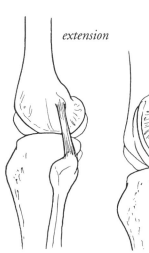

The sensory cells constantly
transmit such information to the
brain, which in turn sends signals to
the muscles via motor neurons. This
is called the **proprioceptive sense.**

Nevertheless, excessive
movement of the joint can
lead to ligamentous stretching
to the point of straining
or tearing the ligament.
This is called a **sprain.**

Muscles

Essentially all movements of the human body result from contraction of muscles. In this book we are concerned with external movements, and will therefore focus on the skeletal muscles (also known as voluntary or striated muscles) which attach to bones. We will not discuss smooth muscle or cardiac muscle.

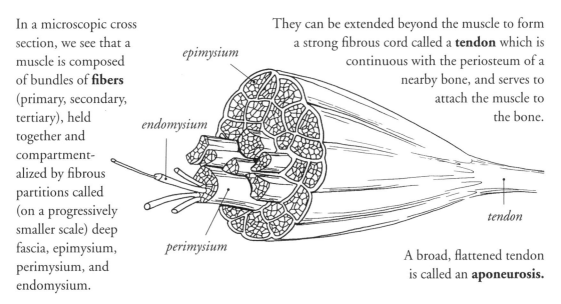

In a microscopic cross section, we see that a muscle is composed of bundles of **fibers** (primary, secondary, tertiary), held together and compartmentalized by fibrous partitions called (on a progressively smaller scale) deep fascia, epimysium, perimysium, and endomysium.

epimysium

endomysium

perimysium

They can be extended beyond the muscle to form a strong fibrous cord called a **tendon** which is continuous with the periosteum of a nearby bone, and serves to attach the muscle to the bone.

tendon

A broad, flattened tendon is called an **aponeurosis.**

These connective tissue partitions (which are continuous with each other) allow easy movement of one muscle or muscle group relative to another.

Each muscle cell, also called a muscle fiber, contains extremely long **myofibrils**. Each of these contains a contractile element in its central part, the sarcostyle. It is striated with dark and light colored bundles arranged in alternating fashion. The structure of these bundles appears (under extreme magnification) to consist of **filaments**:

- dark bundles, thick filaments, which are rounded in the middle (consisting of myosin, a type of protein)

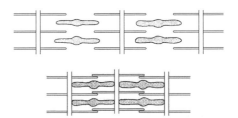

- light bundles, thin filaments, linked to each other through their central part, consisting of actin (another type of protein).

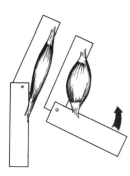

At rest, the actin and myosin filaments are not connected. When the muscle contracts, they unite, and pull each other. This causes a thickening (diameter) and shortening (length), which makes it possible for the muscle to pull the bones to which it is attached.

Typically, a muscle is attached to two different bones. For a given body movement, one bone (called the **origin**) is fixed in some way, and the other (called the **insertion**) moves as a result of muscle contraction. The origin is often the proximal bone, and the insertion the distal bone, but there are many exceptions.

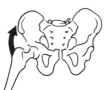

For example, the gluteus medius connects the ilium (large hip bone) to the femur. If the ilium is fixed, contraction of this muscle results in abduction of the femur.

On the other hand, if one is standing with the weight on the leg, the femur becomes the fixed point, and contraction of the muscle results in lateral flexion of the pelvis.

Muscle elasticity

In this book, we will describe muscle actions where the proximal attachment is the fixed point. For some muscles or regions, we will add the muscle action where the distal attachment is the fixed point. Besides their (active) ability to contract, muscles have a (passive) property of elasticity. When stretched, they tend to return to their normal resting length.

For example, the anterior neck muscles, when they contract, are flexors of the neck. During extension of the neck, these muscles become stretched.

When this happens, because of their elasticity, they tend to return the head to its anatomical position.

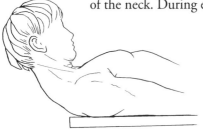

Muscle shapes

Muscles attach to bones in several ways:

- directly via muscle fibers (usually in a broad insertion)
  Example: subscapularis (p. 126)

- via an aponeurosis (broad tendon)
  Example: quadratus lumborum (p. 93)

- via a regular tendon
  Example: coracobrachialis (p. 129)

- via the tendon passing under a fibrous band.
  Example: tibialis anterior (p. 286)

Some muscles have several origins (called "heads"), which may be on more than one bone.

> *Example:* biceps brachii has two heads (p. 147), triceps brachii has three heads (p. 148), and quadriceps femoris has four heads (p. 238)

Typically, the proximal insertion of the muscle is called origin and the distal one is called insertion.

> *Example:* the psoas muscle (p. 92) has its origin on the vertebrae and its insertion on the femur

A muscle can have several origins,

> *Example:* flexor digitorum superficialis originates from both the radius and ulna (p. 177)

…and several insertions.

> *Example:* the interosseous muscles insert on the phalanges and extensor tendons of fingers (p. 180)

The fiber bundles of muscles are arranged in many shapes:

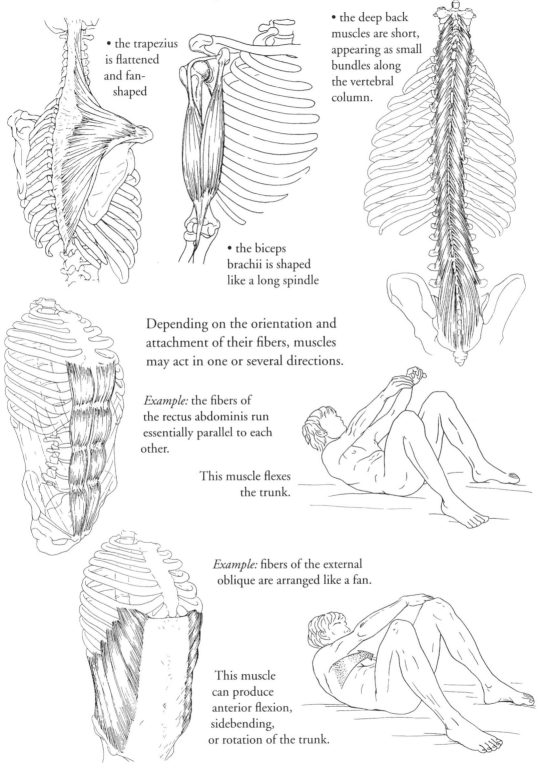

• the trapezius is flattened and fan-shaped

• the deep back muscles are short, appearing as small bundles along the vertebral column.

• the biceps brachii is shaped like a long spindle

Depending on the orientation and attachment of their fibers, muscles may act in one or several directions.

Example: the fibers of the rectus abdominis run essentially parallel to each other.

This muscle flexes the trunk.

Example: fibers of the external oblique are arranged like a fan.

This muscle can produce anterior flexion, sidebending, or rotation of the trunk.

Long muscles are usually kinetic, i.e., able to produce highly visible external motion. Short, deep muscles (e.g., those inserting on the vertebrae or foot bones) tend to be responsible for precise, small-scale adjustments rather than gross movements.

A muscle which crosses and affects a single joint is called **monoarticular**.
A muscle which crosses (and moves) more than one joint is called **polyarticular**.

To stretch a muscle, you move it in a manner that is the opposite
of its usual action around each of its joints.

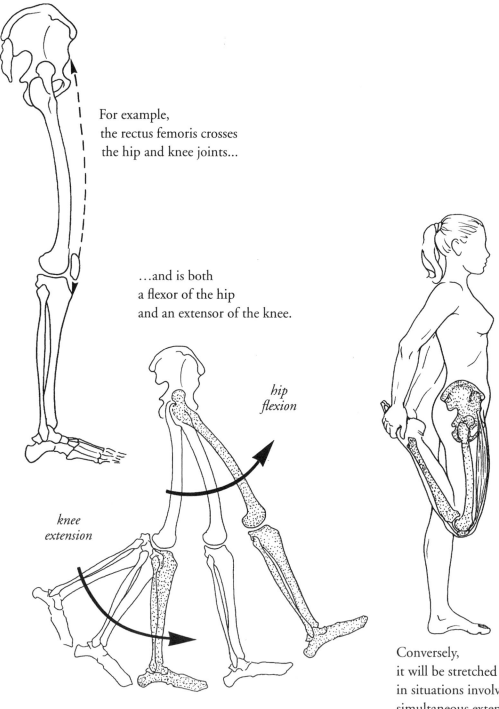

For example,
the rectus femoris crosses
the hip and knee joints...

...and is both
a flexor of the hip
and an extensor of the knee.

*hip
flexion*

*knee
extension*

Conversely,
it will be stretched
in situations involving
simultaneous extension
of the hip and flexion
of the knee.

When we speak of a particular movement, the muscle which produces it is called an **agonist**, and the muscle which produces the opposite movement is called an **antagonist.**

Example: in the case of hip flexion, the psoas major is the agonist...

...and the gluteus maximus (a hip extensor) is the antagonist

Mutually opposing muscles often function together to fix or stabilize a bone.

Example: the serratus anterior and rhomboids have opposite actions; they protract and retract the scapula, respectively, i.e., move it away from and toward the vertebral column. By contracting at the same time, these two muscles work together to fix the scapula.

Different muscles which cooperate to produce the same action are called **synergetic.**

Example: in dorsiflexion of the ankle, three muscles work synergetically: tibialis anterior, extensor hallucis longus, and extensor digitorum longus

When a muscle contracts, it tends to draw its origin and insertion points closer together. Anything that opposes this tendency is called **resistance**.

For example, the brachialis and biceps brachii are the major flexors of the elbow. Their action can be opposed by several types of resistance:

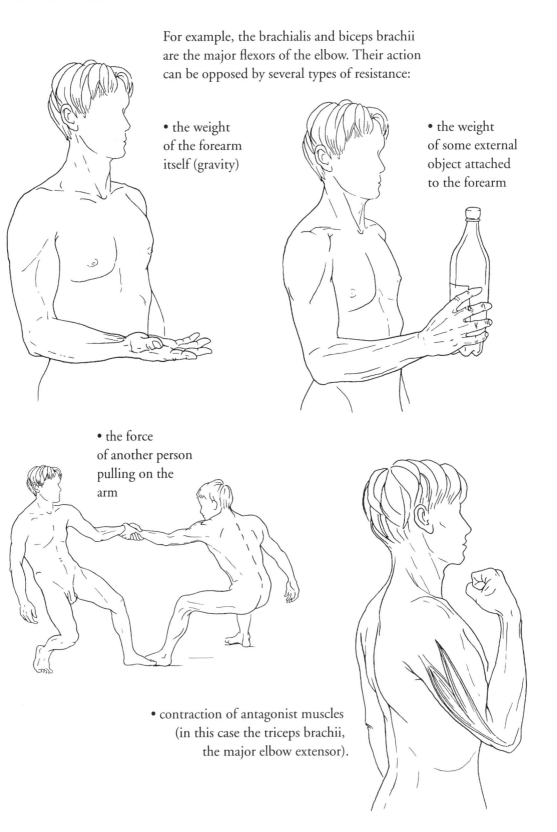

• the weight of the forearm itself (gravity)

• the weight of some external object attached to the forearm

• the force of another person pulling on the arm

• contraction of antagonist muscles (in this case the triceps brachii, the major elbow extensor).

When a muscle contracts, a movement occurs. However, the movement may be caused by forces other than the muscle itself.

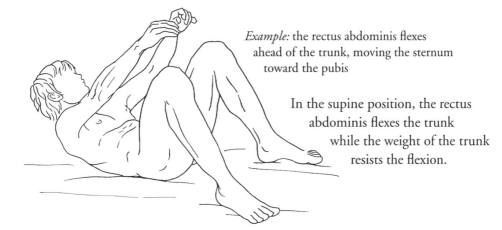

Example: the rectus abdominis flexes ahead of the trunk, moving the sternum toward the pubis

In the supine position, the rectus abdominis flexes the trunk while the weight of the trunk resists the flexion.

However, in a standing position, the rectus abdominis is not active. Instead, gravity causes the trunk to fall forward.

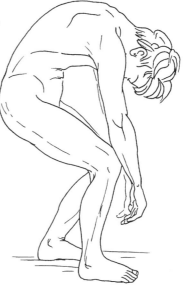

When a movement occurs because of an active muscle, the muscle insertions are moved closer together. This type of contraction is called **concentric**.

In the example of the supine position above, a concentric contraction of the trunk flexors occurs.

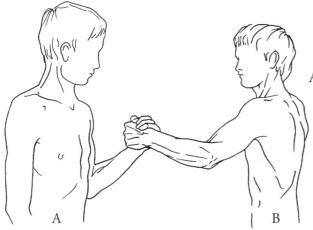

Another example: two people (A and B) pull each other (elbows are flexed). As we can see, A wins. This is a **concentric contraction** of A's elbow flexors.

A   B

In some cases, a muscle works without initiating the action itself; instead, it "applies the brakes" to the action. Without the work of this muscle braking the action, it would occur faster.

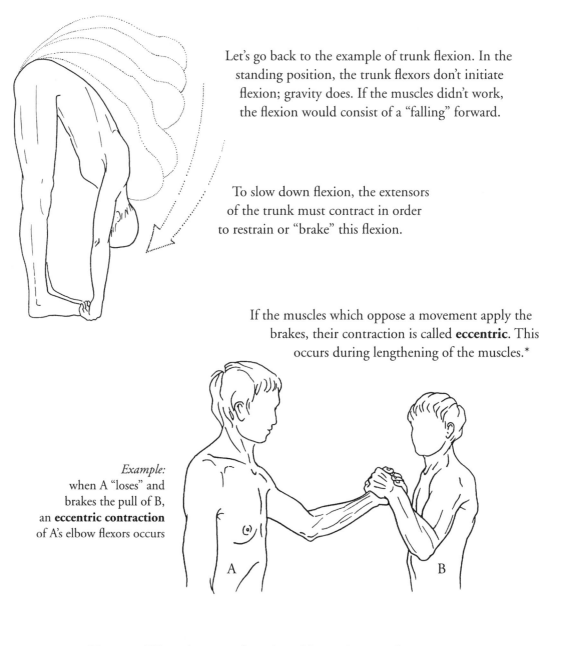

Let's go back to the example of trunk flexion. In the standing position, the trunk flexors don't initiate flexion; gravity does. If the muscles didn't work, the flexion would consist of a "falling" forward.

To slow down flexion, the extensors of the trunk must contract in order to restrain or "brake" this flexion.

If the muscles which oppose a movement apply the brakes, their contraction is called **eccentric**. This occurs during lengthening of the muscles.*

Example:
when A "loses" and brakes the pull of B, an **eccentric contraction** of A's elbow flexors occurs

A   B

Exception: When the rectus femoris and hamstring muscles combine their actions by flexing the hip and knee (e.g., when squatting or doing a "grand plié"), the bones involved change their position, but the muscles do not lengthen or shorten. The hip and knees "cancel out" each other's actions.

It is also possible that a muscle is contracted even though no movement is taking place.

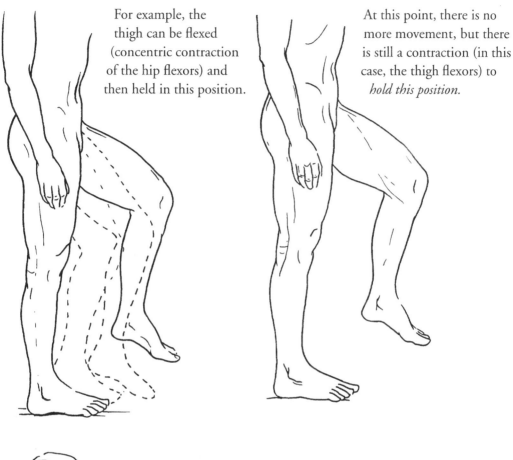

For example, the thigh can be flexed (concentric contraction of the hip flexors) and then held in this position.

At this point, there is no more movement, but there is still a contraction (in this case, the thigh flexors) to *hold this position.*

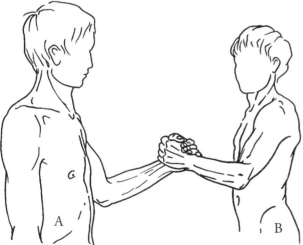

When a position is fixed by a muscular contraction, we call this contraction **isometric**. *The insertions of the muscles do not move.*

A and B are in balance: This is an **isometric contraction**.

In real life, these different types of contraction are combined during various movements.

For example, if we apply the illustration above to the knee, the following would happen when stretching the knee: isometric contraction at the hip flexors and isotonic (concentric) contraction of the knee extensors.

CHAPTER TWO

The Trunk

. . . .

The trunk is the central part of the body. In this book, we will examine only its locomotor functions, not its internal organs.

The trunk serves a double function, which is connected to its bony structure, the vertebral column.

On one hand, the trunk can bend and perform curved movements, like those of a serpent or a measuring tape (unlike the limbs, which perform angular movements like those of a folding measuring stick). This mobility of the trunk is due to the flexibility of the vertebral column, which has twenty-six "levels" of articulation.

On the other hand, the vertebral column contains a tunnel for the nerves: the spinal cord and the nerve roots, which exit the spinal cord. Weakness in the vertebrae may therefore affect not only the joints, but also the spinal cord and nerves as well. Thus, the trunk must be able to align and stabilize the vertebral segments when the body is motionless, and especially when it is carrying a load.

This dual function depends on a finely integrated system of mostly polyarticular muscles, which are either deep (composed of numerous small bundles) or superficial (usually arranged like broad sheets).

Movements of the pelvis are difficult to separate from those of the vertebral column. Therefore it too will be included in this chapter.

Landmarks

Some visible and palpable landmarks of the trunk are shown below.

[FRONT VIEW]

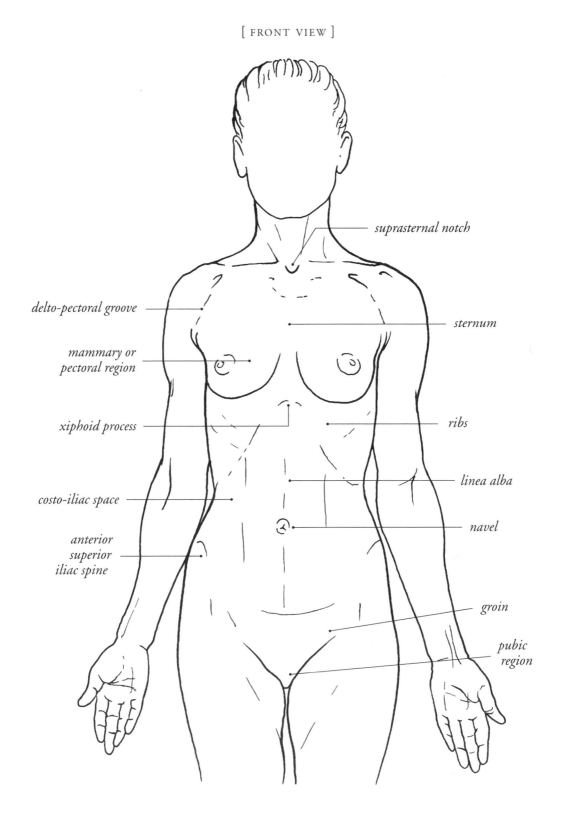

suprasternal notch

delto-pectoral groove

sternum

mammary or
pectoral region

xiphoid process

ribs

linea alba

costo-iliac space

navel

anterior
superior
iliac spine

groin

pubic
region

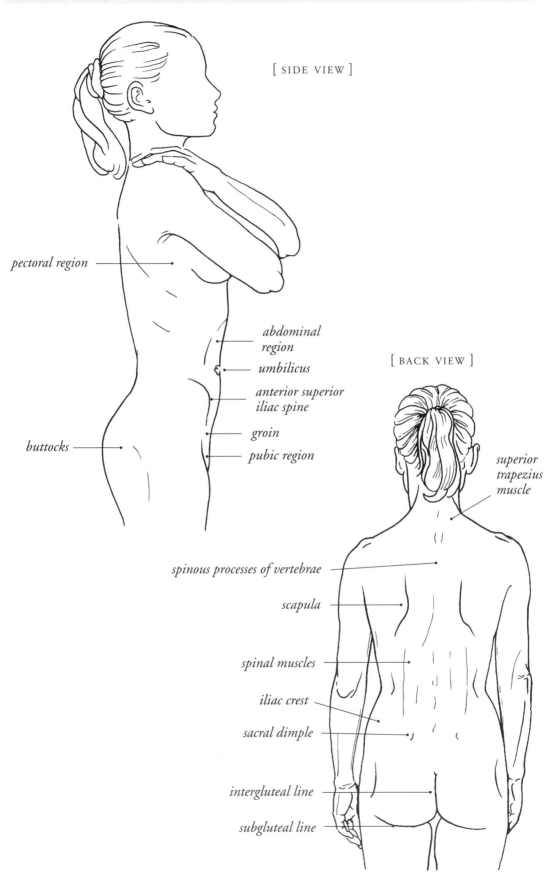

[SIDE VIEW]

pectoral region

abdominal
region

umbilicus

[BACK VIEW]

anterior superior
iliac spine

groin

buttocks

pubic region

superior
trapezius
muscle

spinous processes of vertebrae

scapula

spinal muscles

iliac crest

sacral dimple

intergluteal line

subgluteal line

Movements of the trunk

Thanks to the mobility of the vertebral column, the trunk
can move in several directions, as seen on pages 8-10.

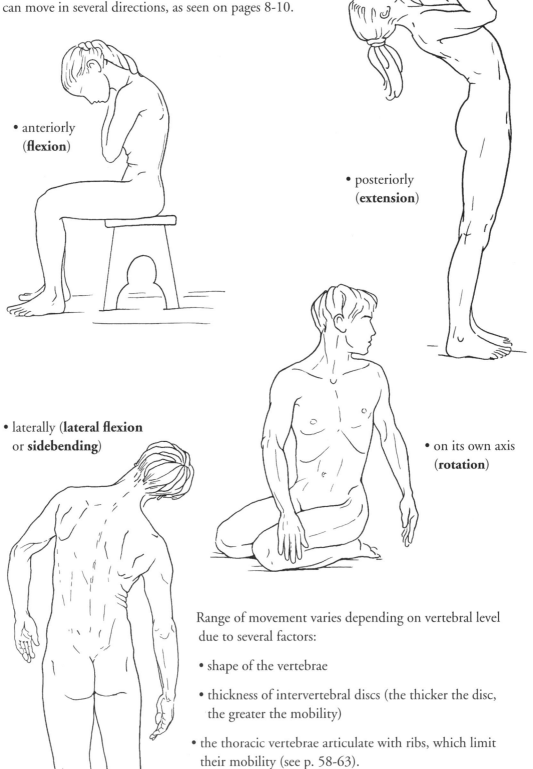

- anteriorly
  (**flexion**)

- posteriorly
  (**extension**)

- laterally (**lateral flexion**
  or **sidebending**)

- on its own axis
  (**rotation**)

Range of movement varies depending on vertebral level
due to several factors:

- shape of the vertebrae

- thickness of intervertebral discs (the thicker the disc,
  the greater the mobility)

- the thoracic vertebrae articulate with ribs, which limit
  their mobility (see p. 58-63).

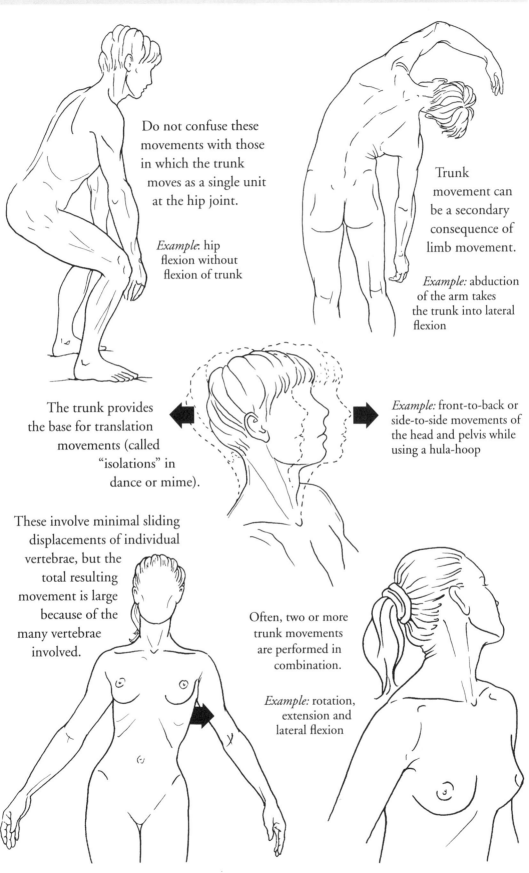

Do not confuse these movements with those in which the trunk moves as a single unit at the hip joint.

Example: hip flexion without flexion of trunk

Trunk movement can be a secondary consequence of limb movement.

Example: abduction of the arm takes the trunk into lateral flexion

The trunk provides the base for translation movements (called "isolations" in dance or mime).

Example: front-to-back or side-to-side movements of the head and pelvis while using a hula-hoop

These involve minimal sliding displacements of individual vertebrae, but the total resulting movement is large because of the many vertebrae involved.

Often, two or more trunk movements are performed in combination.

Example: rotation, extension and lateral flexion

Vertebral column (or spine)

The spine forms a mobile bony stem which constitutes a part of the skeleton of the **trunk**. From top to bottom, it consists of several areas:

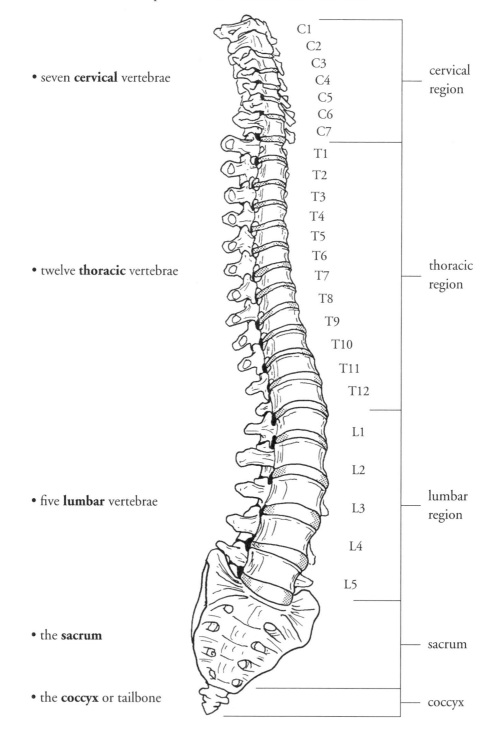

• seven **cervical** vertebrae

• twelve **thoracic** vertebrae

• five **lumbar** vertebrae

• the **sacrum**

• the **coccyx** or tailbone

C1
C2
C3
C4
C5
C6
C7

T1
T2
T3
T4
T5
T6
T7
T8
T9
T10
T11
T12

L1
L2
L3
L4
L5

cervical region

thoracic region

lumbar region

sacrum

coccyx

Within each region, vertebrae are numbered sequentially from top to bottom. For convenience, we usually refer to them by a letter plus a number. *Examples:* C7 = seventh cervical vertebra; T3 = third thoracic vertebra; L2 = second lumbar vertebra; S1 = first sacral vertebra, etc.

There are several characteristic curvatures of the vertebral column:

- sacrum, convex toward the back

- concave lumbar region (the term **lordosis** can refer either to an exaggeration of this curvature, or to the normal condition)

- convex thoracic region (**kyphosis**)

- concave cervical region.

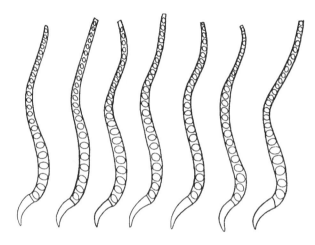

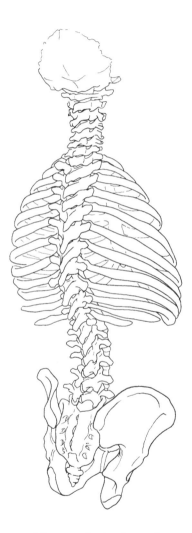

Exact form of these curvatures varies from person to person; this is normal. For example, kyphosis is almost non-existent in some individuals.

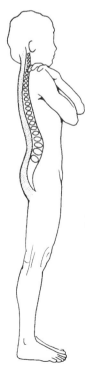

External appearance of these curvatures can be affected by overlying "soft" structures.

Example: a person with large buttocks may appear to have more pronounced lordosis than a person with small buttocks. However, X-rays could reveal identical lumbar curvatures, as shown in these two illustrations.

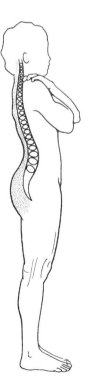

The vertebral column is attached to other skeletal structures: the base of the cranium, the ribs, and the pelvis (ilium).

Vertebral structure

Each vertebra consists of two main parts: the massive **body** (anterior), and the **vertebral arch** (posterior).

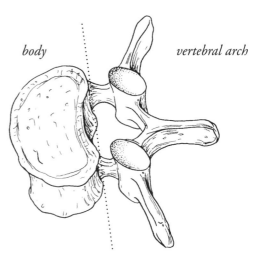

body   *vertebral arch*

This page shows a typical vertebra. Depending on its position in the spine, its shape and size will vary (see also p. 54-71).

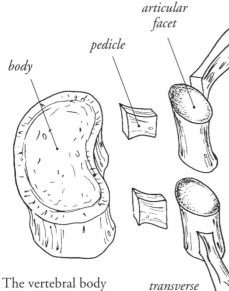

body

pedicle

articular facet

transverse process

The vertebral arch can be divided into many parts. It is connected to the body by two **pedicles.** Two **lamina** unite posteriorly to form a **spinous process.**

lamina

spinous process

The vertebral body is roughly cylindrical and consists of six surfaces.

The thickened junctions between the pedicles and laminae have superior and inferior cartilaginous **articular facets** and a laterally-projecting **transverse process.**

The vertebral holes stacked on top of each other form a bony pipe. This is the spinal canal, through which the spinal cord passes.

vertebral canal

intervertebral foramina

The opening between the body and the arch is called the vertebral foramen. As foramina of many vertebrae are lined up, they form the **vertebral canal** through which the **spinal cord** passes.

The spaces between the pedicles of adjacent vertebrae form a series of openings called **intervertebral foramina.** As spinal nerves branch off the spinal cord, they exit through these foramina.

Each vertebra is attached to its neighbor by three joints (except for the atlas-axis joint; see p. 70).

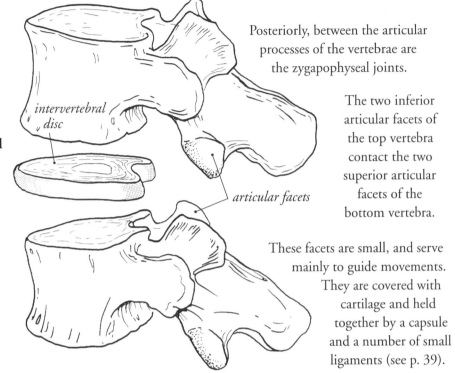

Anteriorly, the bodies are joined by the fibro-cartilaginous **intervertebral disc**.

intervertebral disc

Posteriorly, between the articular processes of the vertebrae are the zygapophyseal joints.

The two inferior articular facets of the top vertebra contact the two superior articular facets of the bottom vertebra.

articular facets

These facets are small, and serve mainly to guide movements. They are covered with cartilage and held together by a capsule and a number of small ligaments (see p. 39).

In cross section, we can see that the disc contains two distinct types of material:

The peripheral area, called the **annulus fibrosus,** is composed of concentric rings of fibrocartilage arranged like the layers of an onion.

annulus fibrosus

nucleus pulposus

The central **nucleus pulposus** is made of a gelatinous substance. The disc, besides allowing movement between vertebrae, acts as a shock absorber and weight bearer.

Ligaments of the spinal column

There are three ligaments extending the length of the vertebral column.

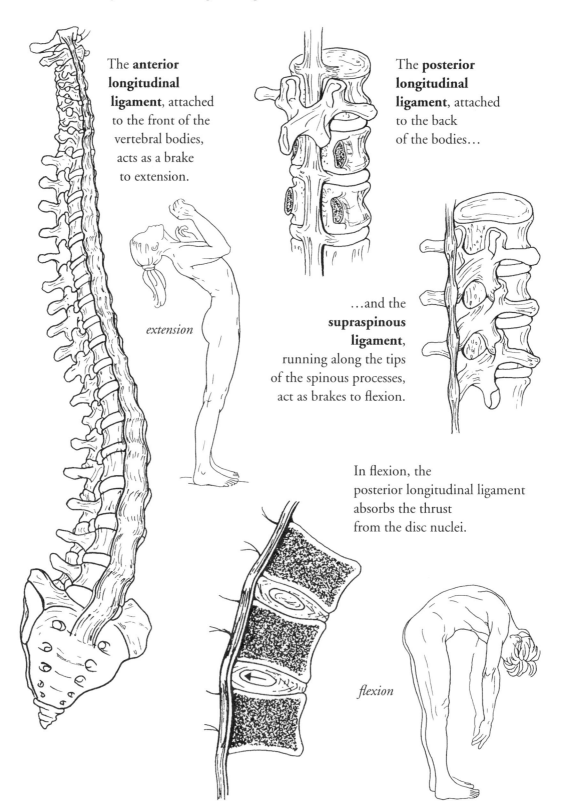

The **anterior longitudinal ligament**, attached to the front of the vertebral bodies, acts as a brake to extension.

extension

The **posterior longitudinal ligament**, attached to the back of the bodies…

…and the **supraspinous ligament**, running along the tips of the spinous processes, act as brakes to flexion.

In flexion, the posterior longitudinal ligament absorbs the thrust from the disc nuclei.

flexion

The other ligaments are discontinuous and hold the posterior arches of the individual vertebrae together.

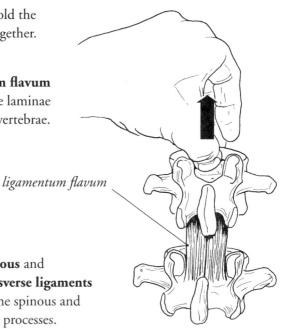

The **ligamentum flavum** connect the laminae of adjacent vertebrae.

ligamentum flavum

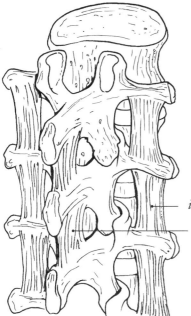

Interspinous and **intertransverse ligaments** connect the spinous and transverse processes.

intertransverse ligament

interspinous ligament

These ligaments are elastic, and can be pierced by a syringe during a spinal tap.

The surfaces of the articular processes are linked together through a capsule inserted on its circumference. The inside of this capsule is reinforced by an extension of the ligamentum flavum and, at the back, by a posterior ligament.

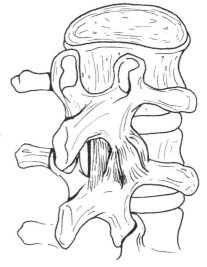

Left sidebending stretches the right intertransverse ligaments.

Other ligaments, specific to certain regions, will be mentioned later.

Movements of the vertebrae

We can think of the vertebral column as a series of fixed segments (the vertebrae) having mobile connections (discs, ligaments).

In **flexion**, B tilts toward the front. The superior articular facets slide on the inferior ones.

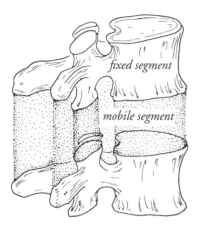

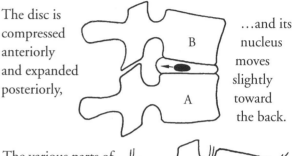

The disc is compressed anteriorly and expanded posteriorly,

...and its nucleus moves slightly toward the back.

The various parts of the vertebral arches are pulled apart, and the ligaments connecting these parts are stretched.

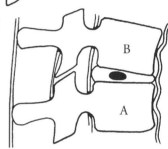

Movements of individual vertebrae are compounded such that the entire structure has considerable mobility in three dimensions.

Type and extent of mobility varies with different spinal regions, depending on size and shape of vertebrae, and other factors.

Let's look at what happens between two vertebrae during movement, assuming that the top vertebra (B) is mobile, while the bottom vertebra (A) is fixed.

In **extension**, the opposite occurs. B tilts toward the back.

The disc is compressed posteriorly and expanded anteriorly,

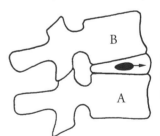

...and its nucleus moves forward.

The articular facets are pressed together, the arches move closer together,

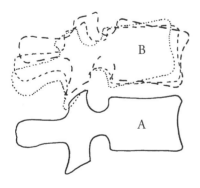

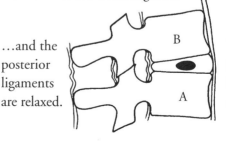

...and the posterior ligaments are relaxed.

The anterior longitudinal ligament is stretched.

What happens in **lateral flexion**? Let's consider left sidebending as an example.

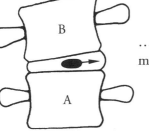

The left sides of A and B move closer together,

...while the right sides move farther apart.

The disc is expanded (and its nucleus moves) to the right side.

left

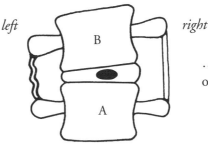

right

On the left side, the transverse processes and articular facets come closer together, and the associated ligaments are relaxed;

...the opposite occurs on the right side.

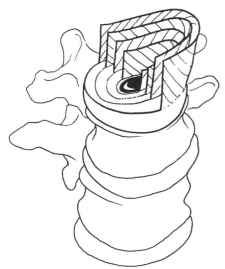

In **rotation**, the fibers of the disc, whose orientation alternates from one layer to the next, are under torsion (twisting).

They are moving in two different directions from layer to layer, such that one layer is stretched and the next layer is relaxed.

Because of the effect of the torsion, and the ensuing tension on the fibers, the distance between the vertebrae is diminished. The nucleus is therefore slightly compressed.

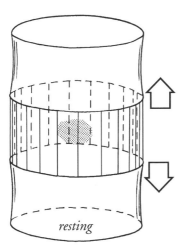

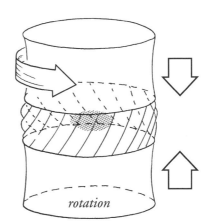

All of the connecting ligaments are stretched by rotation.

resting

rotation

The intervertebral disc serves as shock absorber

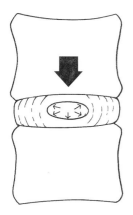

The disc often receives pressure from the vertebral body above. The nucleus, because of its central location and gelatinous composition, tends to distribute the pressure it receives in every direction.

Thus, the fibers of the annulus receive both horizontal and vertical pressures.

As long as fluid remains in the nucleus, the disc performs its role as a shock absorber very efficiently.

Unfortunately, due to the aging process and/or excessive wear and tear, the disc may partially lose this property, i.e., cracks develop in the annulus through which the fluid of the nucleus can escape.

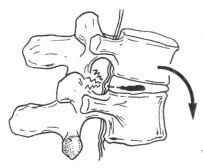

This condition is termed a herniated or **ruptured disc**. It happens most commonly as a result of chronic flexion movements, during which the nucleus moves toward the back and fluid can escape there. The fluid may then compress the nerve roots, e.g., the sciatic nerve which exits from the lumbar region, where pressures on the vertebral column are most intense.

This situation, combined with chronic or sudden extreme tension on the posterior longitudinal ligament, can result in chronic lumbar backaches.

To avoid these problems, it is important to avoid "loaded" vertebral flexion, e.g., flexing the lumbar spine while lifting a heavy object.

In fact, it is preferable to avoid loaded lumbar flexions in any type of physical exercise, even if you are not lifting any object.

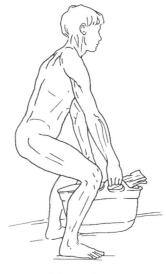

Instead, keep the spine straight and flex at the hip and knee joints only.

Pelvis (or pelvic girdle)

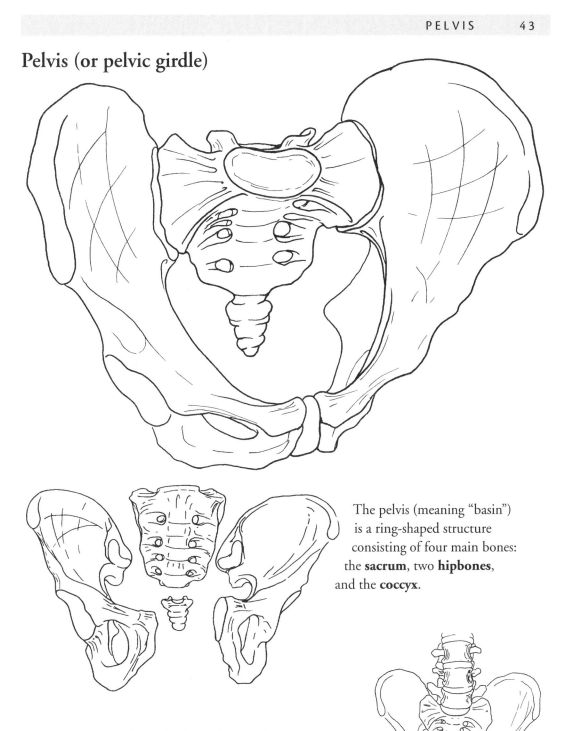

The pelvis (meaning "basin") is a ring-shaped structure consisting of four main bones: the **sacrum**, two **hipbones**, and the **coccyx**.

There are associated muscles (e.g., those making up the pelvic floor) and ligaments above.

The pelvis receives the weight of the upper body and passes this weight on to the lower limbs via its articulations with the femurs.

Conversely, it must absorb stresses from the lower limbs, e.g., in walking or jumping.

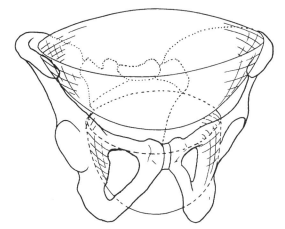

The shape of the bones forms
a **greater (false) pelvis** at the top
and a **lesser (true) pelvis** at the bottom.

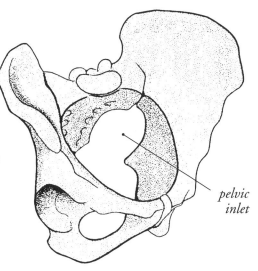

*pelvic
inlet*

The top opening of the lesser pelvis
is called the **pelvic inlet**.

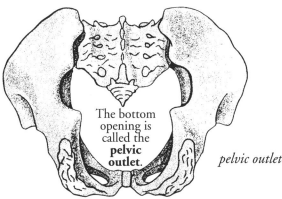

The bottom
opening is
called the
**pelvic
outlet**.

pelvic outlet

The hip bones of the pelvis

The hip bones are flat and consist of a
superior and inferior part, which are twisted
against each other (a little bit like a propeller).
In the adult, each hip consists of three bones
that are fused together: the **ilium**, **ischium**,
and **pubis**.

These bones fuse at a cartilage in the shape
of a Y, located in the center of the socket.

The hip has two surfaces (medial and lateral)
and four edges (superior, inferior, anterior,
and posterior).

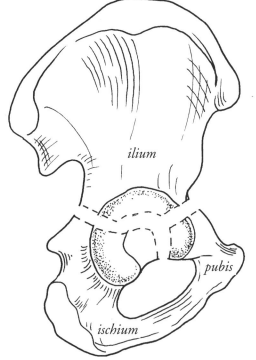

ilium

pubis

ischium

On the lateral surface
of the hip bone, we find:

• the superior border
or **iliac crest**

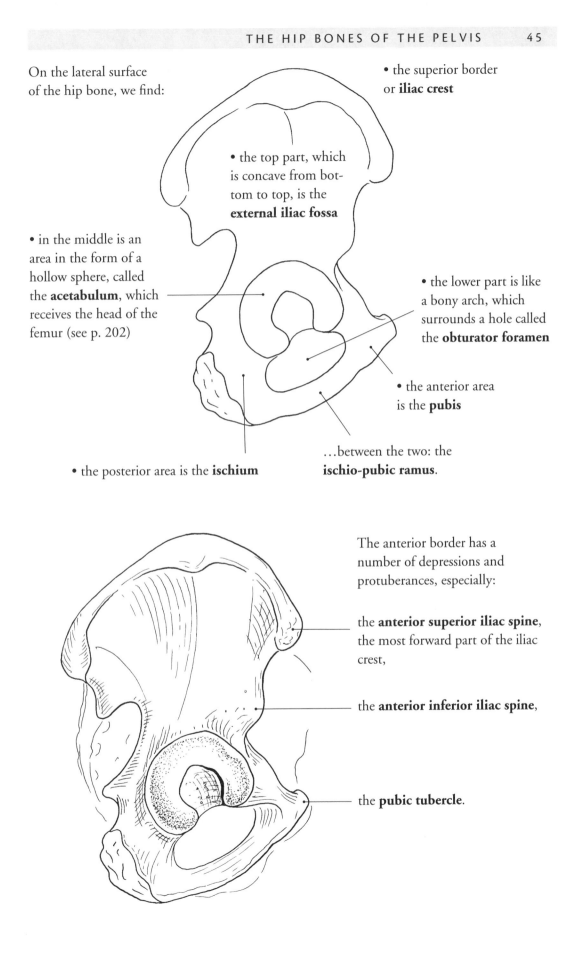

• the top part, which
is concave from bot-
tom to top, is the
external iliac fossa

• in the middle is an
area in the form of a
hollow sphere, called
the **acetabulum**, which
receives the head of the
femur (see p. 202)

• the lower part is like
a bony arch, which
surrounds a hole called
the **obturator foramen**

• the anterior area
is the **pubis**

• the posterior area is the **ischium**

...between the two: the
ischio-pubic ramus.

The anterior border has a
number of depressions and
protuberances, especially:

the **anterior superior iliac spine**,
the most forward part of the iliac
crest,

the **anterior inferior iliac spine**,

the **pubic tubercle**.

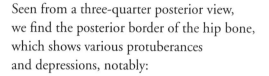

Seen from a three-quarter posterior view,
we find the posterior border of the hip bone,
which shows various protuberances
and depressions, notably:

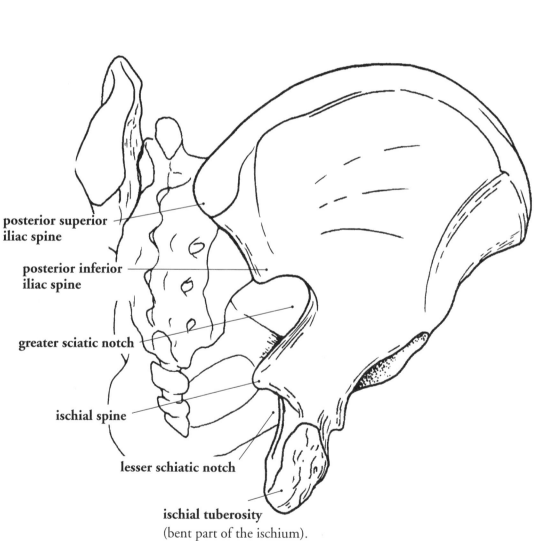

**posterior superior
iliac spine**

**posterior inferior
iliac spine**

greater sciatic notch

ischial spine

lesser schiatic notch

ischial tuberosity
(bent part of the ischium).
This is the bone on which you sit.

On the medial surface, we find:

the **internal iliac fossa**

an oblique crest,
the **iliopectineal line,**
which forms the border
between the lesser and
greater pelvis

the internal circumference
of the obturator foramen

an articular surface located in front of the pubis,
in the shape of an oval covered by cartilage,
which unites the two pubic bones.
This is the pubic symphyseal surface.

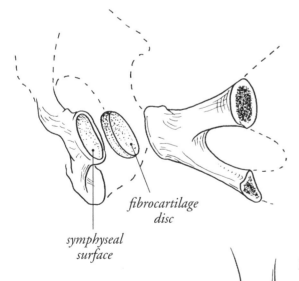

*fibrocartilage
disc*

*symphyseal
surface*

The articulation between the
two pubic bones is called the
pubic symphysis.

Between the two surfaces is
a fibrocartilage disc which
attaches to the articular
surfaces.

The entire structure is covered by
a fibrous cuff, reinforced by four
ligaments: anterior, posterior,
superior, and inferior.

This is a joint with very little mobility.
Only small gliding, compression, and
twisting movements are possible here.
During childbirth, the joint loosens so that
the pelvis can open up further.

The shape and proportions of the pelvis vary considerably in normal individuals.

For example, the pelvic inlet may have a round shape,

...or be compressed in either direction (center, right).

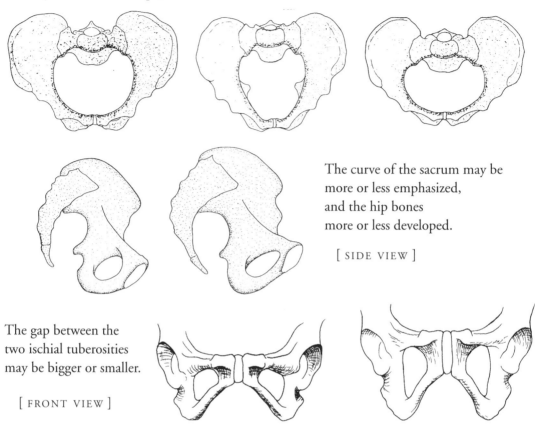

The curve of the sacrum may be more or less emphasized, and the hip bones more or less developed.

[SIDE VIEW]

The gap between the two ischial tuberosities may be bigger or smaller.

[FRONT VIEW]

The sacral crest or posterior superior iliac spines may protrude in some individuals. This condition, combined with lack of "padding" by muscles and adipose tissue, can result in difficulty doing floor exercises, particularly rolling on a hard surface.

There are obvious differences in pelvic shape between males and females.

Essentially, the male pelvis is narrower and the female pelvis is wider, with larger pelvic inlet and outlet.

male pelvis

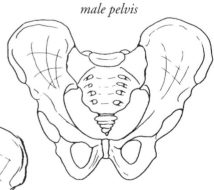

female pelvis

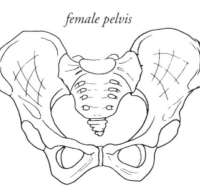

These differences are all related to the ability of women to carry and deliver a child.

The pelvis is sometimes also called the pelvic girdle. Anatomically, girdles are bony and articular structures by which the limbs are attached to the trunk.

The two girdles

At the top of the trunk, lying on top of the ribs, is the **shoulder girdle**, which consists of the sternum, the two clavicles, and the two scapulae.

Through this structure, the upper limbs are attached to the trunk.

Its main feature is its mobility. It is not linked via joints to the spinal column, but instead is linked to the thoracic cage.

(For more information, see p. 110-115.)

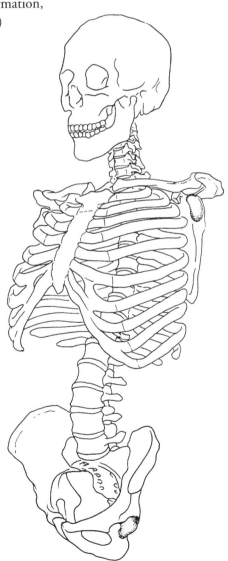

At the bottom of the trunk is the **pelvic girdle**, or pelvis, which consists of the sacrum and the two hip bones.

Through this structure, the lower limbs are attached to the trunk. Because the articulations between these bones are not very mobile, it is a very stable structure. The pelvic girdle is attached to the trunk via the sacro-lumbar joint, which connects it to the spinal column.

(For more information, see p. 43-53 in this chapter.)

Sacrum

The sacrum is the posterior, wedge-shaped component of the pelvic ring, located between the two ilia. It is composed of five expanded, fused vertebrae (S1-S5) whose individual components are readily visible.

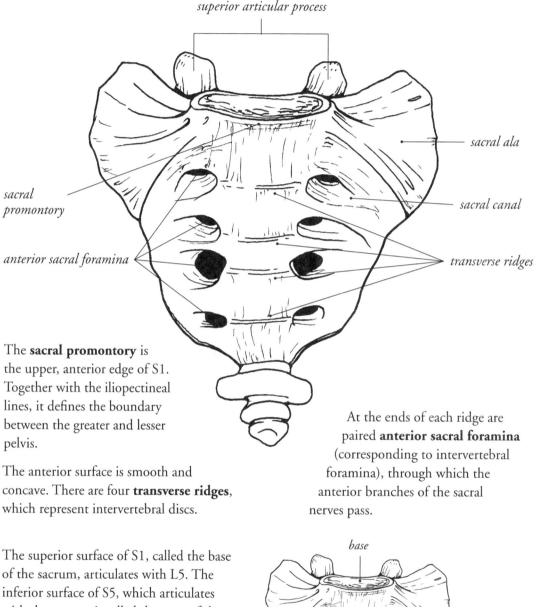

The **sacral promontory** is the upper, anterior edge of S1. Together with the iliopectineal lines, it defines the boundary between the greater and lesser pelvis.

The anterior surface is smooth and concave. There are four **transverse ridges**, which represent intervertebral discs.

At the ends of each ridge are paired **anterior sacral foramina** (corresponding to intervertebral foramina), through which the anterior branches of the sacral nerves pass.

The superior surface of S1, called the base of the sacrum, articulates with L5. The inferior surface of S5, which articulates with the coccyx, is called the apex of the sacrum.

sacral canal

The convex posterior surface is much rougher.

The **posterior sacral foramina** are continuous with the anterior foramina; posterior branches of the sacral nerves exit here. The **median sacral crest** and paired **lateral sacral crests** represent the spinous and transverse processes of the vertebrae. The spinal cord enters through the **sacral canal**.

fusion of the sacral lamina

median sacral crest

lateral sacral crest

posterior sacral foramina

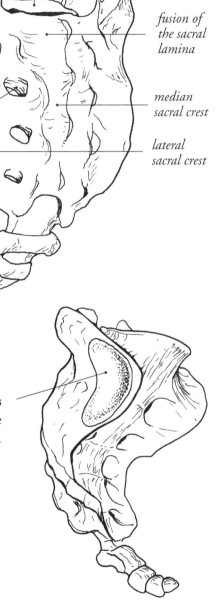

The **auricular** ("ear-shaped") **surfaces** are best seen in side view. These articulate with the auricular surfaces of the ilia.

superior cornua

transverse process

Coccyx

The coccyx is a small triangular bone consisting of the fusion of three to five vertebrae. (It is not possible to tell them apart.) The coccyx articulates with the sacrum via an **oval-shaped surface**. It is supported by a capsule and ligaments (this joint is often fused).

Sacroiliac joint

This joint consists of the two auricular surfaces
on top of the ilium and the sacrum.

The auricular surface of the hip bone
is slightly convex…

The auricular surface
of the sacrum is
slightly concave.

*auricular
surfaces*

…especially
in the lower
portion.

This arrangement
allows for an im-
portant movement
between the sacrum
and the two ilia,
called nutation
and counternutation.

In **nutation**, the sacral base
tilts anteroinferiorly,

…and the sacral apex
tilts posterosuperiorly.

In one type
of movement,
which we shall call
"adduction" of the
pelvis, the sacrum
goes into nutation,
the iliac "wings"
are pulled medially,

…and the ischial tuberosities
move laterally.

Thus, the sacral promontory
moves toward the pubis,
and the sacral apex
moves away from the pubis,

…while the ischial bones
move away from each other,
increasing the distance between them.

In summary:
during adduction,
the diameter of the
pelvic outlet increases and
the pelvic inlet decreases
from front to back.

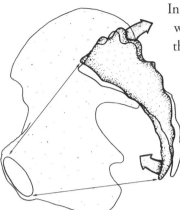

In the opposite movement, which we will call "abduction" of the pelvis, the sacrum moves in **counternutation**, as the sacral base tilts backward and upward and the sacral apex tilts forward and downward. The iliac wings move away from the midline and the ischial bones move toward each other.

In summary: during abduction, the pelvic inlet becomes larger from front to back, and the diameter of the pelvic outlet becomes smaller.

The dimensional changes between the pelvic inlet and outlet occur especially during childbirth: At the beginning of the baby's engagement in the pelvis, there will be a sacral counternutation, and when the baby comes out (expulsion), a nutation.

Sacroiliac ligaments

Each sacroiliac joint is reinforced by a capsule and a strong network of ligaments:

Anteriorly, there are two fasciae (not shown).

Inferiorly, the **sacrospinous** and **sacrotuberous ligaments** connect the sacrum to the ischial spine and ischial tuberosity respectively.

These ligaments tend to oppose "adduction" of the pelvis.

sacrospinous ligament

sacrotuberous ligament

There are also a series of **posterior sacroiliac ligaments** connecting the ilium to the lateral sacral crest.

These tend to oppose "abduction" of the pelvis.

posterior sacroiliac ligaments

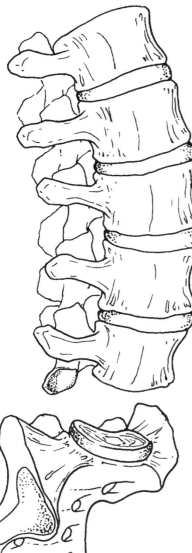

Lumbar spine

This is the next structure above the sacrum.
It is concave at the back.
This is the "loin" area
between the pelvis and the ribcage.

Lumbar vertebrae

The lumbar vertebrae are large,
and become larger
at the bottom of the lumbar spine.

The discs are thick,
about one-third the thickness
of the vertebral bodies.
This increases mobility.

The bodies of the vertebrae are large,
and shaped like a lima bean.
They are concave posteriorly.

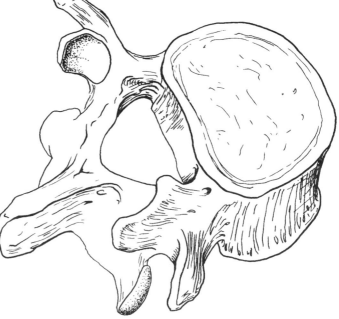

The transverse processes
are long and have distinct
tubercles on the end for
muscle attachment.

The articular processes project
above and below the vertebral body,
with a narrow
part in-between:
the isthmus.

The superior articular processes
of the vertebrae have a concave
cylindrical form.
They point medially
(and slightly backward).

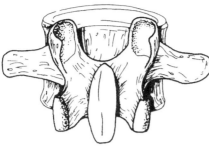

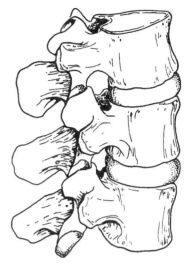

The inferior articular processes have a convex cylindrical form.
They point laterally (and slightly forward).
These bony projections articulate with the neighboring
vertebrae and are stacked on top of each other.

Their articular facets are vertical and
somewhat sagittal.* These characteristics
make rotation very difficult (below),

…but facilitate
movements
of flexion,

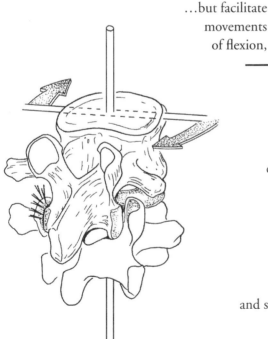

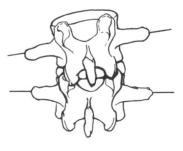

extension,

and sidebending.

*The top lumbar vertebrae are sagittal.
They become more and more frontal
towards the lower lumbar vertebrae.
At the lumbosacral junction,
they are completely frontal.

To summarize the mobility of the lumbar vertebrae:
good range of motion for flexion, extension,
and sidebending, limited ROM for rotation.

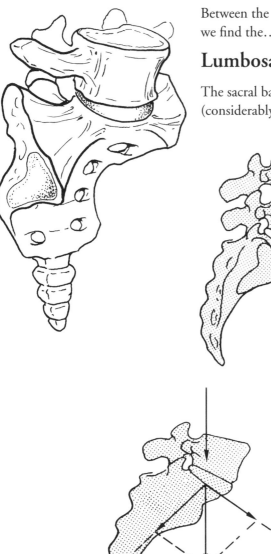

Between the sacrum and the 5th lumbar vertebra we find the...

Lumbosacral joint

The sacral base is tilted forward to a variable degree (considerably, in some individuals).

L4

L5

S1

Also, the body of L5 and the disc between L5 and S1 are slightly thicker anteriorly than posteriorly (this also applies to the L4/L5 joint).

Thus, the lumbosacral joint is concave posteriorly.

Due to the oblique orientation of the sacrum, there are two perpendicular forces resulting from the weight of the upper body at L5.

One is directed along the axis of the sacrum,

...while the other tends to make L5 slide forward.

This second force can be significant if the sacral base is greatly tilted (right). In other words, L5 does not really "rest" on the sacral base, as many people think. It "wants" to slide forward.

Notice that this tendency is opposed by the contact between the articular facets of S1 and the inferior articular processes of L5.

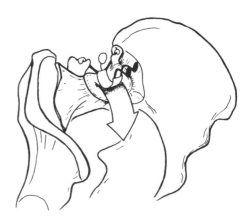

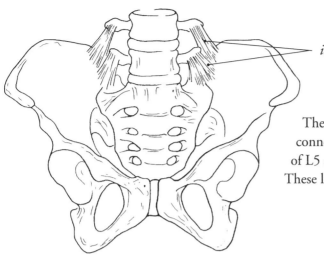

iliolumbar ligaments

There are **iliolumbar ligaments**
connecting the transverse processes
of L5 and L4 to the iliac crest.
These ligaments tend to oppose sidebending.

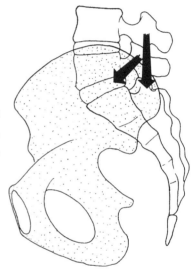

From the side, you can see that the
ligament from L4 is oriented posteroinferiorly,
while that from L5 runs more anteroinferiorly.

Thus, the L4 ligament is stretched
and the L5 ligament relaxed
during flexion…

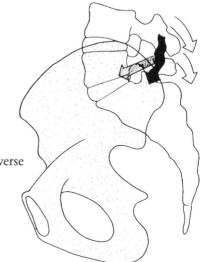

…while the reverse
occurs during
extension.

Thoracic spine

The thoracic spine articulates with the ribs.
It consists of twelve vertebrae,
the thoracic vertebrae.

In contrast to the lumbar region, the thickness
of the discs is only about one-sixth that of the
bodies; this tends to limit range of motion.

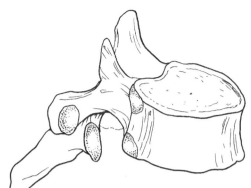

The bodies of the thoracic vertebrae
are cylindrical, and almost round
in cross section.

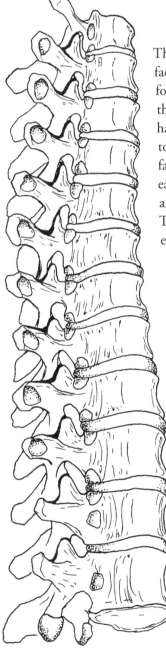

There are posterior
facets on the bodies
for articulation with
the heads of ribs. TI
has a facet near the
top, and an inferior
facet. T2 through T9
each have superior
and inferior facets.
T10 through T12
each have one facet.

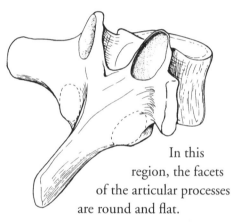

In this
region, the facets
of the articular processes
are round and flat.

The articular facets are oriented postero-superolaterally
on the superior articular processes, and anteroinfero-
medially on the inferior processes. This arrangement
permits flexion, extension, and sidebending.

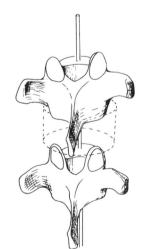

These facets
are located
approximately
on the circum-
ference of a
circle whose
center would
be the center
of the vertebral
body.

This facilitates
rotation.

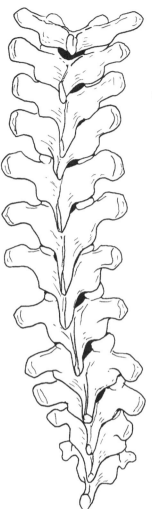

The laminae are flat, rectangular, and taller than they are wide. They are stacked on and overlap each other like tiles on a roof.

The transverse processes decrease in length from top to bottom (left), and those of T1 through T10 have anterior facets for articulation with the tubercles of ribs.

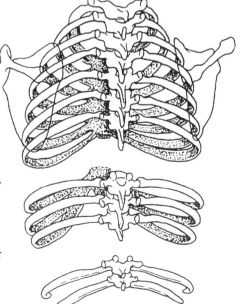

The spinous processes are elongated, laterally compressed, and (again in contrast to the lumbar region) directed inferiorly (except for T11 and T12). These characteristics help prevent hyperextension.

The vertebral attachment to the rib cage tends to limit mobility of the thoracic spine.

This is particularly true for TI through T7 between the scapulae, whose corresponding ribs (the "true ribs") are connected directly to the sternum by short pieces of cartilage which allow little mobility.

Ribs 8, 9, and 10 (the "false" ribs) have longer costal cartilages which attach to the cartilage of rib 7 rather than directly to the sternum. The mobility of T8, T9, and T10 is correspondingly greater.

Ribs 11 and 12 (the "floating" ribs) have no anterior attachment at all, so the mobility of T11 and T12 is greatest.

Thoracic cage

The thoracic cage consists of the thoracic vertebrae at the back, and the ribs and sternum in the front.

clavicular notch   *jugular notch*

manubrium

The **sternum** is a flat bone located at the front of the thorax. It consists of three parts:

1. manubrium: The top part of the manubrium articulates with the clavicles (see p. 110).

2. body: The lateral edge has another 7 notches, which articulate with the first 7 costal cartilages.

3. xiphoid process (not always present).

body

xiphoid process

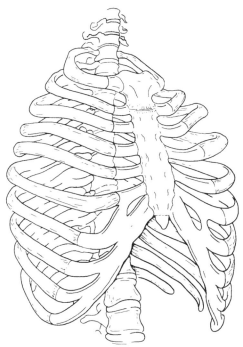

The **ribs** are elongated, flattened, and twisted bones. Each rib consists of a posterior end with three parts: head, neck and tubercle; a body; and an anterior end which articulates with the costal cartilage.

head

neck

tubercle

The first rib is the smallest. It is flattened from top to bottom.

body

anterior end

The rib is curved in three different ways:

A. from above, the edge of the rib has the shape of a bucket handle

B. from the front, it has the shape of an italic *S*

C. when we "straighten" part of a rib, we can see torsion along its long axis.

A

B

C

The rib is like a curved blade "under tension." (During a sternotomy, a surgical procedure where the sternum is cut, one can see how the ribs move apart.)

Most ribs articulate with two thoracic vertebrae at three points, as noted above: the two facets on the head of the rib contact the vertebral bodies, and the tubercle contacts the transverse process.

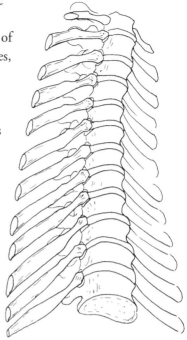

Exceptions are ribs 1, 11 and 12, which only contact one vertebral body, and ribs 11 and 12, which do not rest on a transverse process.

Each joint is stabilized by numerous small ligaments.

This illustration depicts the joints as if pulled apart from each other.

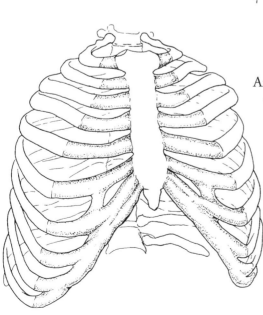

Anteriorly, each rib is attached to the sternum via its **costal cartilage**, which increases the elasticity of the thoracic cage.

The first seven ribs are short and attach directly to the sternum. These ribs are called **true ribs**.

The three following cartilages, which are longer, all attach to the 7th rib. These are called **false ribs**, and have greater mobility.

The two lowest ribs do not have a cartilage. These are called **floating ribs**.

Movements of the ribs

The movements of a rib can be compared
to those of a bucket handle.

As the rib moves up or down,
the diameter of the thoracic cage
("bucket") is changed.

Posteriorly, the rib pivots
on an axis passing through
the center of two joints:

- one, with two facets,
  articulates with the vertebral body
- another articulates with the
  transverse process of the vertebra.

At different levels, the spatial relationship of these two
joints changes because the shape of the vertebra changes.

For the superior
thoracic vertebrae,
the axis is directed
more laterally,
and elevation of the rib increases the
anterior diameter of the thoracic cage.

By contrast,
for the inferior
thoracic vertebrae,
the axis is directed
more posteriorly,
and elevation of the rib
increases the lateral diameter
of the thoracic cage.

The
anterior ends
of the ribs easily
accommodate these
movements because of the
flexibility of the costal cartilages,
which varies depending on the level of the
rib. However, this flexibility can decrease
with age, thus reducing overall mobility.

During **costal inhalation,** the ribs are elevated. Therefore, the diameter of the upper thoracic cage is increased in an anterior direction, while that of the lower cage is increased in a lateral direction.

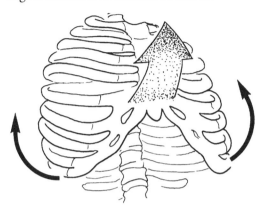

The reverse occurs during **costal exhalation,** when the ribs move down. The costal cartilages undergo some torsion (twisting on their own axis) on inhalation, then return to their normal shape on exhalation.

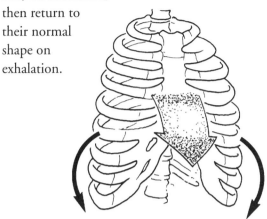

Obviously, spinal movements also affect the orientation of the ribs:

The ribs in front move closer together during thoracic flexion,

...and farther apart during extension.

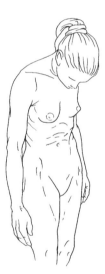

In right sidebending, the ribs move closer together on the right side and farther apart on the left side.

In right rotation, the ribs move posteriorly on the right and anteriorly on the left.

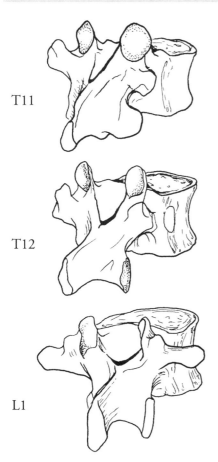

T11

T12

L1

Thoracolumbar junction

The junction between the thoracic
and lumbar spinal regions
has some interesting features.

T12 resembles T11 in its upper half, but resembles L1
in its lower half, i.e., the spinous process is shortened
(which facilitates extension), and the inferior articular
facets are large and convex (which restricts rotation).

At T12/L1, the junction between the thoracic
and lumbar spinal regions, mobility of the spine
is characterized by good flexion, extension, and
sidebending, but very little rotation.

Because T11 and T12 are both
attached to floating ribs, and
because of the shape of the spinous
process and inferior articular processes on T11, there is good range of motion
in all directions at the T11/T12 joint: flexion,
extension, sidebending, and rotation.

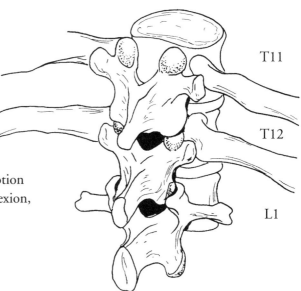

T11

T12

L1

Starting from the bottom of the spine,
the T11/T12 joint is the first
important rotational joint.

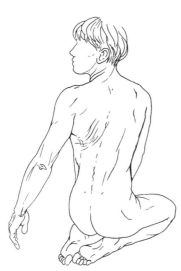

Hyper-rotation is possible
at this location.

Cervical spine

The cervical spine forms the skeleton of the neck.
We will study two regions.

1. The **suboccipital cervical spine**, which consists of the two top vertebrae:

C1, or **atlas**, which is located just below the skull

C2, or **axis**.

These two vertebrae have a different shape and function differently from the other vertebrae, and will be studied separately.

2. The **lower cervical spine**, consisting of C7 through C3. All of these vertebrae have the same characteristics.

Cervical vertebrae

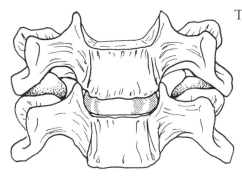

The bodies of cervical vertebrae are small. The discs are about one-third as thick as the bodies. Both of these factors tend to increase mobility.

Sidebending is somewhat restricted by the rectangular shape of the bodies, as seen in cross section.

The superior surfaces of the cervical vertebral body project upward on the sides. (These projections are called uncinate processes.) The inferior surfaces are shaped to articulate with the superior ones.

This bony configuration provides mobility as well as stability. The vertebral bodies are "stacked" laterally.

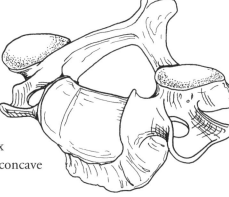

Additionally, the superior surfaces are slightly convex and tilted forward. The inferior surfaces are slightly concave and lifted backward.

The lengths of the spinous processes vary.

Those in the middle cervical area are short, especially C4, which permits good extension of the cervical spine.

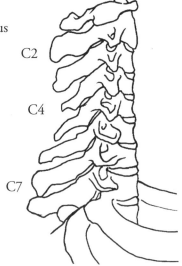

C2

C4

C7

However, at the bottom part of the cervical spine (the area of C6 and C7), extension is more limited because these two spinous processes are long and restrain extension.

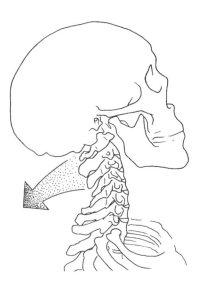

Each of the cervical transverse processes has two roots:

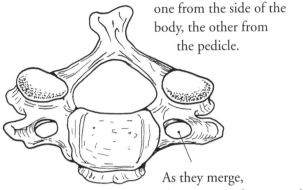

one from the side of the body, the other from the pedicle.

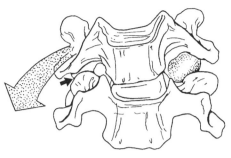

They are thick and tend to limit side-bending when they come in contact.

As they merge, the two roots form a medial opening called the **transverse foramen,** and a small lateral groove.

The **vertebral artery** passes through the transverse foramina of CI-C6. Spinal nerves pass through the lateral grooves.

Thus, correct alignment of the cervical spine is important for protection of these soft tissues.

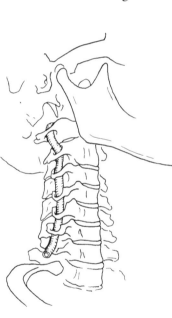

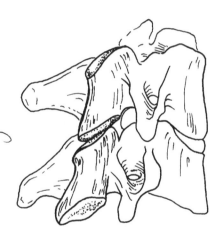

The superior articular facets face posterosuperiorly, and the inferior facets anteroinferiorly, at an angle of 45°.

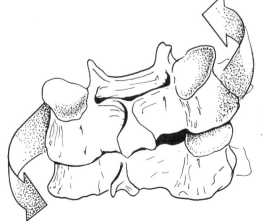

Sidebending of the cervical spine is always accompanied by a certain amount of rotation.

For example, during left sidebending, the left facet moves inferiorly and slightly posteriorly, while the right facet moves superiorly and slightly anteriorly. The combination of these two movements results in slight left rotation.

The **suboccipital spine** is the top part of the cervical spine.

This is the area where movements which are independent of the head are made, like nodding the head "yes" and shaking it "no."

It consists of two unique vertebrae: the atlas and the axis.

The atlas

This is the first vertebra at the top of the spine.

It is not shaped like the other vertebrae. It is essentially a bony ring, reinforced by two strong lateral masses.

In front is the anterior arch (the atlas has no vertebral body).

In back is the posterior arch (the atlas has no spinous process).

The large transverse processes project from the sides. They contain the transverse foramina, through which the vertebral artery passes.

The **transverse ligament of the atlas,** which attaches to the inside of the lateral masses, divides the ring of the atlas into two parts.

The anterior part surrounds the dens of the axis (see p.70)

transverse foramen

The posterior part contains the vertebral foramen, through which the spinal cord passes.

transverse ligament

The upper and lower surfaces of the lateral masses have facets for articulation with the occipital condyle at the top...

dens of the axis

occipital condyle

axis

...and the axis at the bottom.

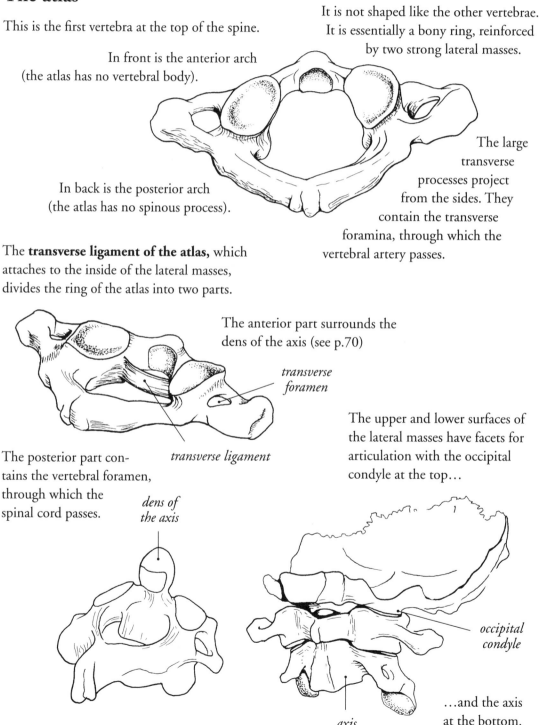

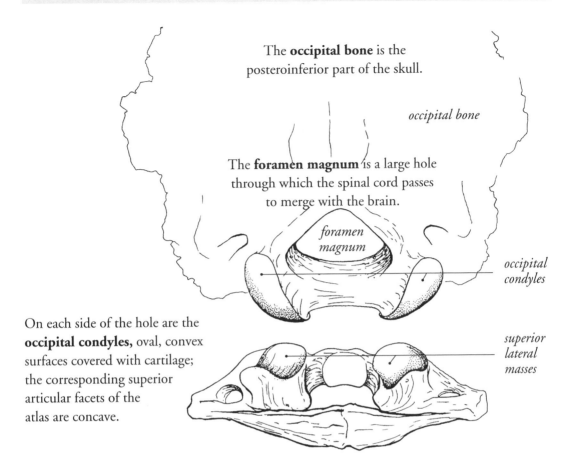

The **occipital bone** is the posteroinferior part of the skull.

occipital bone

The **foramen magnum** is a large hole through which the spinal cord passes to merge with the brain.

foramen magnum

occipital condyles

On each side of the hole are the **occipital condyles,** oval, convex surfaces covered with cartilage; the corresponding superior articular facets of the atlas are concave.

superior lateral masses

These articulating surfaces lie essentially on the outside of an imaginary sphere whose center is inside the skull. Thus, the occipital-atlas joint could be viewed as a ball-and-socket joint, potentially allowing movement in any direction.

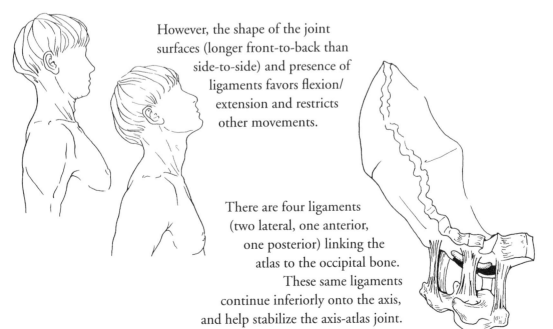

However, the shape of the joint surfaces (longer front-to-back than side-to-side) and presence of ligaments favors flexion/extension and restricts other movements.

There are four ligaments (two lateral, one anterior, one posterior) linking the atlas to the occipital bone. These same ligaments continue inferiorly onto the axis, and help stabilize the axis-atlas joint.

The axis
and its connection with the atlas

The axis is the second cervical vertebra.
It has the shape of a typical cervical vertebra,
except that on the top,
there are certain bony particularities
which help it to articulate with the atlas.

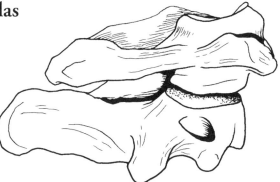

Each side of the vertebral body
has an oval convex surface,
which articulates
with a lateral facet
of the atlas
above.

The vertebral body of the axis
has a peg-like process: this is
the **dens** (odontoid process).
Like a pin, it is lodged
in the anterior part
of the ring-like
structure of
the atlas.

Thus, there is no disc
between the atlas and
axis, just two freely
movable joints (see
p. 14, diarthrosis).

The articulating sur-
faces of atlas and axis
are convex, i.e., they
do not fit into each
other. It is a hinge with
permanent mobility.

There are two articulations
between the atlas and dens:

• The dens fits anteriorly
against the anterior arch of C1,

• and posteriorly against an articular
surface of the transverse ligament of C1.

Thus, the pivot joint of C1-C2
consists of a ring-like structure
covered by cartilage rotating around the dens.

The atlas (C1) presses on the axis (C2) and turns around its pivot.

At this level, rotation is the most important movement (as in shaking your head "no"). This consists of a rotation and a gliding movement:

The axis of rotation can pass either through the dens,

...or one of the articular facets of the atlas and axis.

Rotation of C1 on C2 is accompanied by some lateral gliding of C2 which helps preserve the integrity of the spinal canal.

There are anterior (not shown) and posterior ligaments between C2 and C1, as well as ligaments (shown here as arrows) connecting the occiput to the body and dens of C2.

Due to the convexity of the articular facets of both C1 and C2,

...C1 tends to move slightly closer to C2 during rotation.

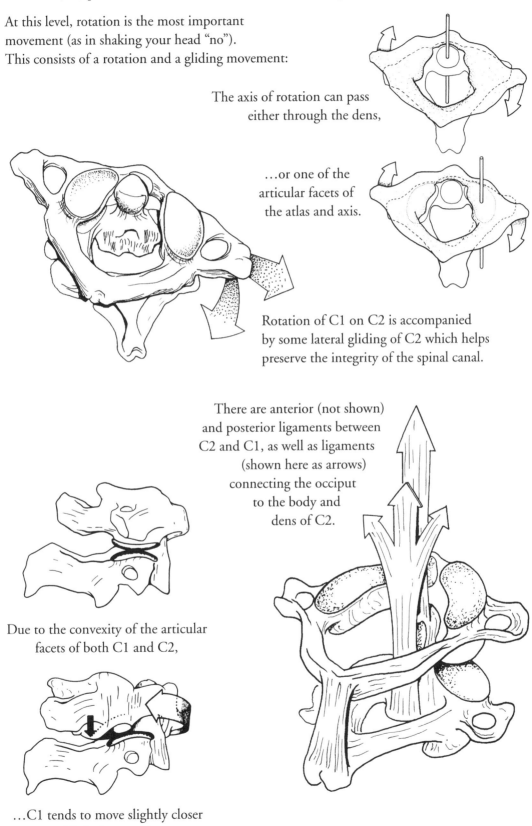

Muscles of the trunk and their bony attachments

Cranial bones:

(mainly occipital and temporal bones)
sternocleidomastoid (SCM)
pre-cervical muscles
suboccipital muscles (deep neck muscles)
semispinalis and longissimus capitis
splenius capitis
trapezius

Ribs:

longissimus thoracis
iliocostalis
serratus posterior
latissimus dorsi
scalenes
intercostal muscles
levatores costarum
transversus thoracis
diaphragm
abdominal muscles

Shoulder girdle, humerus:

levator scapulae
rhomboid
latissimus dorsi
trapezius
sternocleidomastoid (SCM)

Vertebrae:

spinal muscles
splenius
levator scapulae
serratus posterior
rhomboid
latissimus dorsi
trapezius
longus colli
pre-cervical muscles
scalenes
levatores costarum
diaphragm
psoas
quadratus lumborum
abdominal muscles

Pelvic girdle:

muscles of the lumbar spine
latissimus dorsi
psoas
quadratus lumborum
abdominal muscles
muscles of the pelvic floor

Femur:

psoas

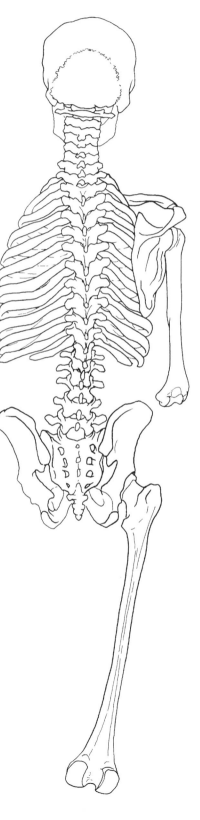

Posterior muscles of trunk and neck

The back side of the trunk contains numerous muscles, which are arranged in several layers. The deepest ones attach only to the vertebrae and consist of small bundles, which pass from one vertebra to the next.

The **intertransverse** muscles connect one transverse process to the next, posterior to the intertransverse ligament.

Action: sidebending

The **interspinalis** muscles connect adjacent spinous processes, on either side of the ligament.

Action: extension

Innervation: posterior branches of spinal nerves (C3-S4)

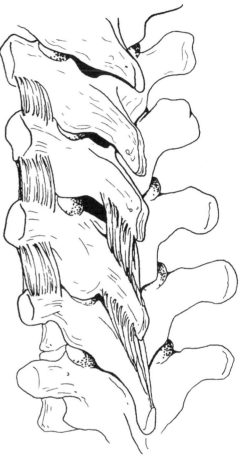

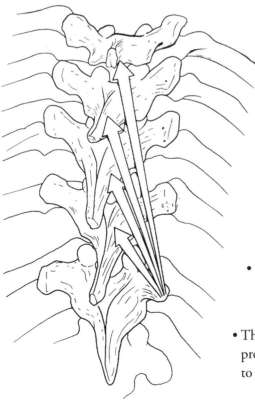

The **transversospinalis** muscles attach to the back of the vertebrae all along the spine, from sacrum to axis. They consist of three bundles, which arise at the transverse processes:

• The **rotatores** pass toward the lamina of the vertebra above.

• The **multifidus** pass to the spinous processes of the vertebrae that are located two to four levels above.

• The **semispinalis** muscles pass to the spinous processes of the vertebrae that are located four to six levels above, covering the other layers.

When viewed from behind,
the transversospinalis form
a "chevron"-like pattern
along the posterior spinal column.

Actions: The
muscle fibers
run diagonally:

• from inferior
to superior:
when contracting
bilaterally,
they move the
vertebrae into
extension

• from medial to
lateral: sidebending

• from anterior to posterior:
rotation of the spine
(opposite the side of
contraction).

Innervation: posterior branches
of spinal nerves (C3-S4)

Role of deep spinal muscles in keeping trunk erect

Muscles located in the convex sagittal curves of the spine elicit a chain reaction.

Electromyographic studies have shown that the activity and importance of the **transversospinalis** muscles are variable depending upon spinal level, particularly in terms of "elongating" (straightening) the spine:

It is important around T6, where the posterior convexity of the thoracic spine is most pronounced.

It is less significant at T12.

It is also less significant around L3, where the posterior concavity of the lumbar region is most pronounced.

As shown here, the actions of the transversospinalis muscles are more predominant where the spine is most convex in the back.

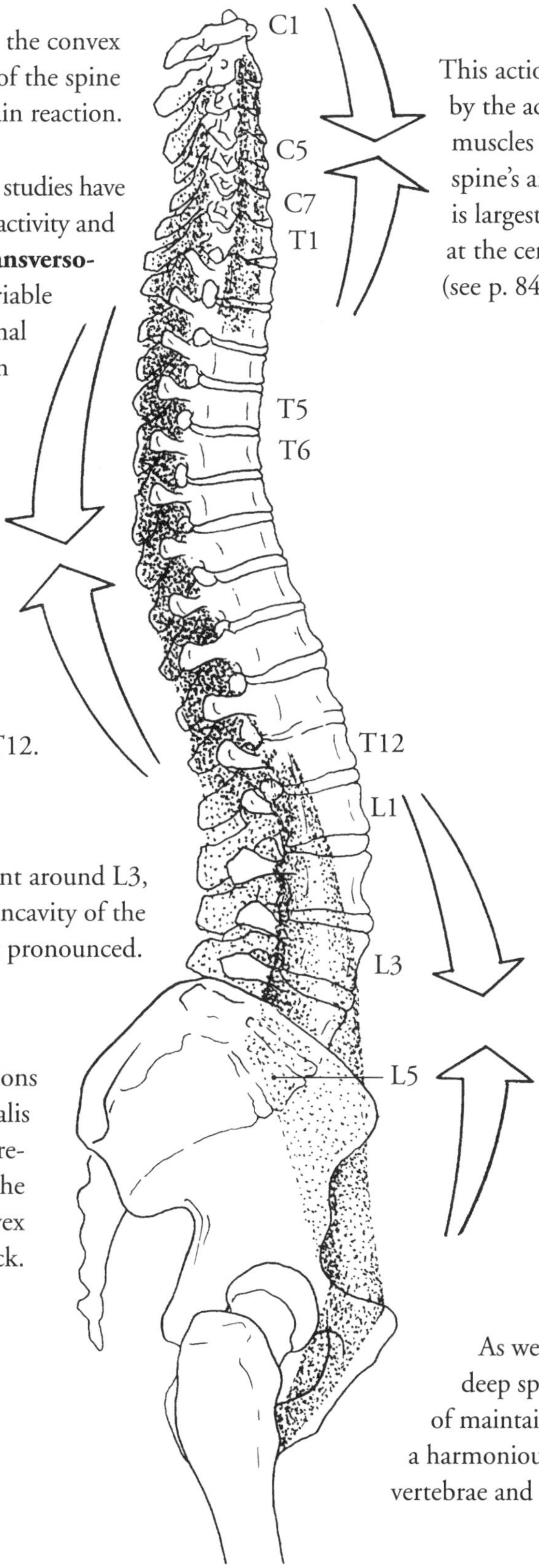

This action is completed by the actions of two other muscles located where the spine's anterior convexity is largest: the **longus colli** at the cervical level (see p. 84)…

… and the **psoas** at the lumbar level (p. 92).

As we can see, this group of deep spinal muscles is capable of maintaining stability and a harmonious alignment of the vertebrae and intervertebral discs.

C1
C5
C7
T1
T5
T6
T12
L1
L3
L5

Deep neck muscles

These are analogous to the transversospinalis
but insert on the occiput.

Rectus capitis posterior minor runs from the
posterior arch of C1 to the inferior occipital ridge.

Rectus capitis posterior major originates
from the spinous process of C2 and inserts
just lateral to the minor.

rectus capitis posterior:

minor

major

Obliquus capitis superior
originates from the transverse process of C1
and inserts on the occiput
lateral to rectus capitis posterior major,
just posterior to the mastoid process
of the temporal bone.

Obliquus capitis inferior
runs from the spinous process of C2
to the transverse process of C1.

Action: extension, sidebending,
and rotation of C1 on C2 (not shown)

Action: these three muscles
help produce extension at
the C1/ C2 joint
when they
contract
bilaterally

obliquus capitis:

superior

inferior

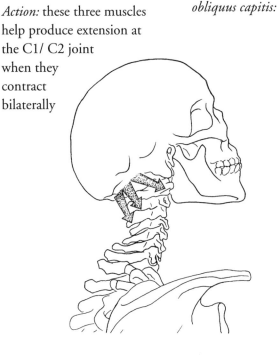

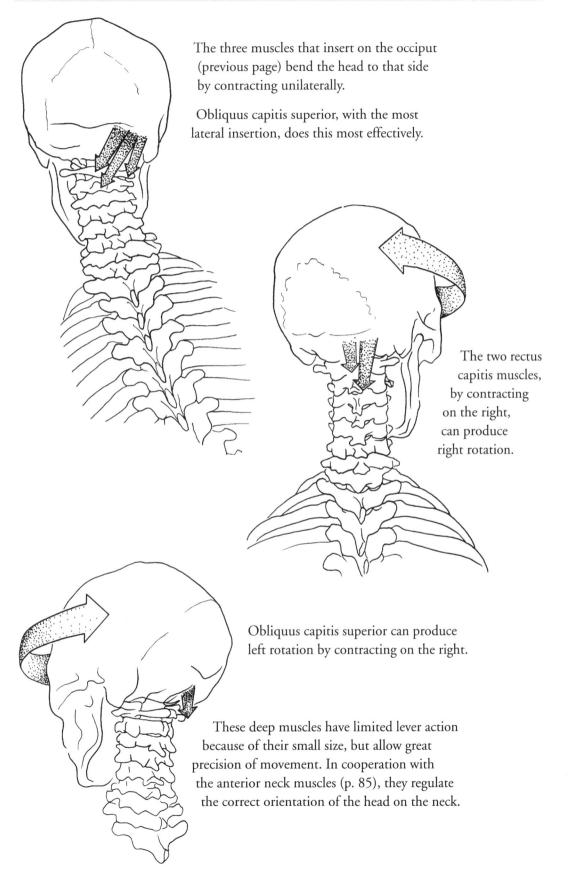

The three muscles that insert on the occiput
(previous page) bend the head to that side
by contracting unilaterally.

Obliquus capitis superior, with the most
lateral insertion, does this most effectively.

The two rectus
capitis muscles,
by contracting
on the right,
can produce
right rotation.

Obliquus capitis superior can produce
left rotation by contracting on the right.

These deep muscles have limited lever action
because of their small size, but allow great
precision of movement. In cooperation with
the anterior neck muscles (p. 85), they regulate
the correct orientation of the head on the neck.

Intermediate back and neck muscles

The group of posterior muscles shown on this page forms a layer superficial to those described on the previous pages. These muscles are sometimes referred to collectively as the sacrospinalis or erector spinae.

There are three muscles in this group: iliocostalis (most lateral), longissimus, and spinalis (most medial; see p. 80). Each of these is further divided into three subcomponents.

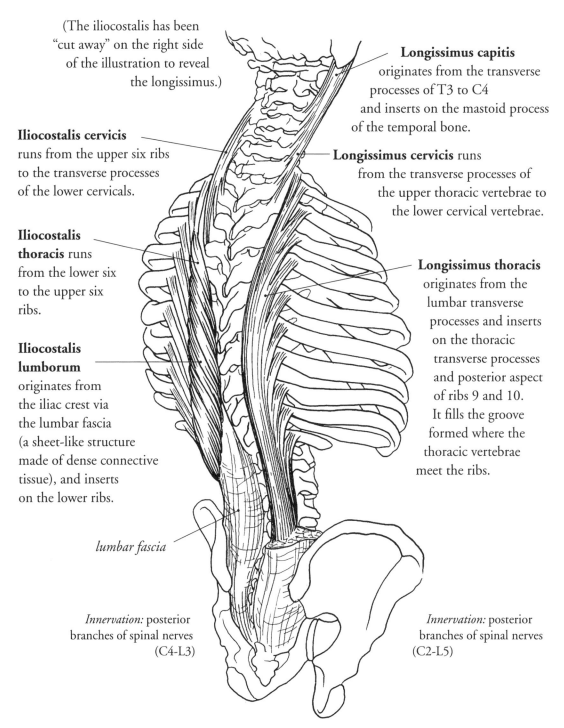

(The iliocostalis has been "cut away" on the right side of the illustration to reveal the longissimus.)

Iliocostalis cervicis runs from the upper six ribs to the transverse processes of the lower cervicals.

Iliocostalis thoracis runs from the lower six to the upper six ribs.

Iliocostalis lumborum originates from the iliac crest via the lumbar fascia (a sheet-like structure made of dense connective tissue), and inserts on the lower ribs.

lumbar fascia

Innervation: posterior branches of spinal nerves (C4-L3)

Longissimus capitis originates from the transverse processes of T3 to C4 and inserts on the mastoid process of the temporal bone.

Longissimus cervicis runs from the transverse processes of the upper thoracic vertebrae to the lower cervical vertebrae.

Longissimus thoracis originates from the lumbar transverse processes and inserts on the thoracic transverse processes and posterior aspect of ribs 9 and 10. It fills the groove formed where the thoracic vertebrae meet the ribs.

Innervation: posterior branches of spinal nerves (C2-L5)

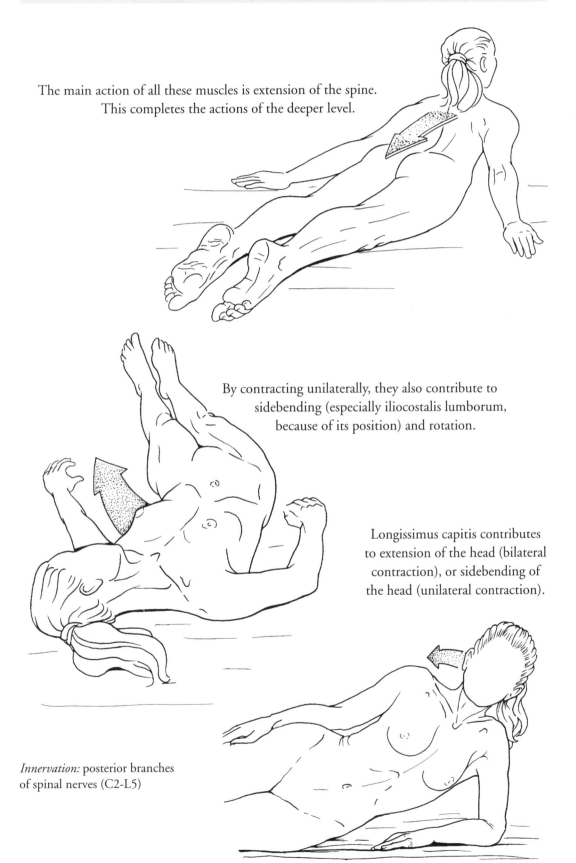

The main action of all these muscles is extension of the spine. This completes the actions of the deeper level.

By contracting unilaterally, they also contribute to sidebending (especially iliocostalis lumborum, because of its position) and rotation.

Longissimus capitis contributes to extension of the head (bilateral contraction), or sidebending of the head (unilateral contraction).

Innervation: posterior branches of spinal nerves (C2-L5)

A second layer of muscles, located along the spinal column, covers the muscles described on the previous pages.

Spinalis capitis and **semispinalis capitis** can be considered together. They originate respectively from the spinous processes of C7–T1 and the transverse processes of C4–T4, and insert on the occiput.

semispinalis capitis; spinalis capitis

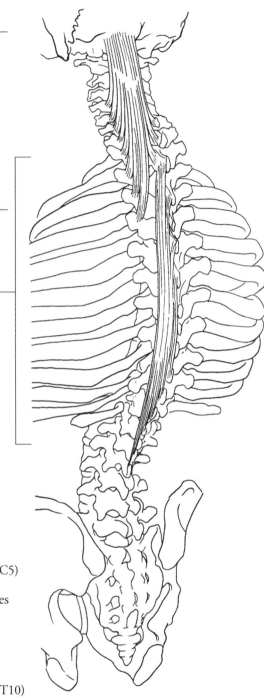

Actions: if they contract bilaterally, while the cervical spine is fixed, they extend the head. If they contract unilaterally, with the spine fixed, they can contribute to sidebending or rotation.

spinalis thoracis

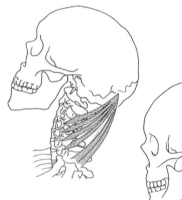

On the other hand, if they contract bilaterally with the head fixed, they tend to straighten out the cervical spine.

Innervation: posterior branches of spinal nerves (C1–C5)

Spinalis thoracis runs from the spinous processes of T1 through T10 to the spinous processes of T11 through L2.

Action: extends the spine in the thoracic region

Innervation: posterior branches of spinal nerves (C2–T10)

The back muscles described thus far (and some later) form a deep muscle layer, also referred to as the **spinal muscles**. They have a short lever arm, hence not much power to do certain activities, e.g., extension of the spine from the horizontal position. However, they work very precisely. In the vertical position, they work together to keep the spine straight, rebalancing each other slightly at every vertebral level. In a standing person, they are almost constantly working. This is because they are very active muscles, able to work without fatiguing for long stretches of time. For example, the head "sits on the neck" for an entire day with the help of these muscles.

Splenius capitis originates from the nuchal ligament and the spinous processes of C7 through T3-T4. It inserts on the mastoid process and adjacent occipital bone.

Splenius cervicis runs from the spinous process of T5-T7 to the transverse processes of C1-C3.

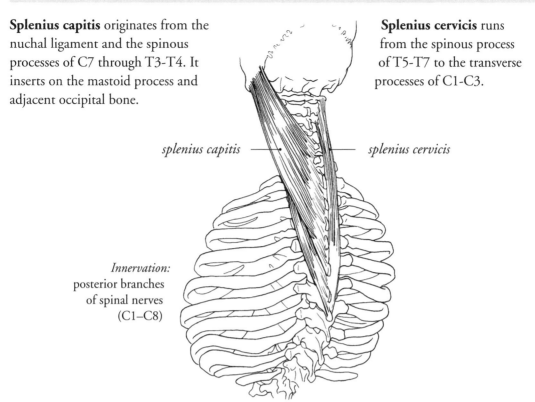

splenius capitis — — *splenius cervicis*

Innervation: posterior branches of spinal nerves (C1–C8)

Actions: contracting bilaterally, these muscles extend the head and cervical spine. Contracting unilaterally, they cause sidebending and rotation toward the contracting side.

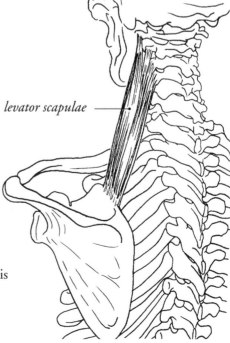

levator scapulae —

Levator scapulae (see p. 123) runs from the transverse processes of C1-C4 to the medial scapula.

Actions: acts primarily on the scapula; however, when the scapula is fixed, its actions can reinforce those of the splenius cervicis

The next layer consists of the **posterior serratus muscles**.

Serratus posterior superior
runs from the spinous processes of C7 to T3
and inserts on the first five ribs.

Action: elevates the ribs
and thereby aids in inspiration

Innervation: branches of first four
intercostal nerves (T1-T4)

Serratus posterior inferior
runs from the spinous processes of T12 to L2
and inserts on the last four ribs.

Action: depresses these ribs
and thereby aids in expiration

Innervation: superior branches of
the last four intercostal nerves

The following three muscles act primarily on the shoulder
joint and will be discussed later in that context. However,
when the shoulder is fixed, they can also act on the spine.

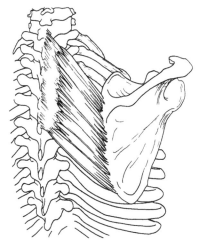

Rhomboids (see p. 123)

Action: when the scapula is fixed, the contraction of these
muscles pulls the vertebrae laterally

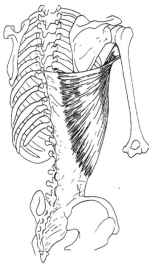

Latissimus dorsi (see p. 131)

Action: when acting bilaterally,
this muscle extends
the thoracolumbar spine

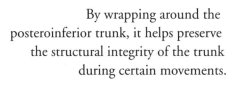

By wrapping around the
posteroinferior trunk, it helps preserve
the structural integrity of the trunk
during certain movements.

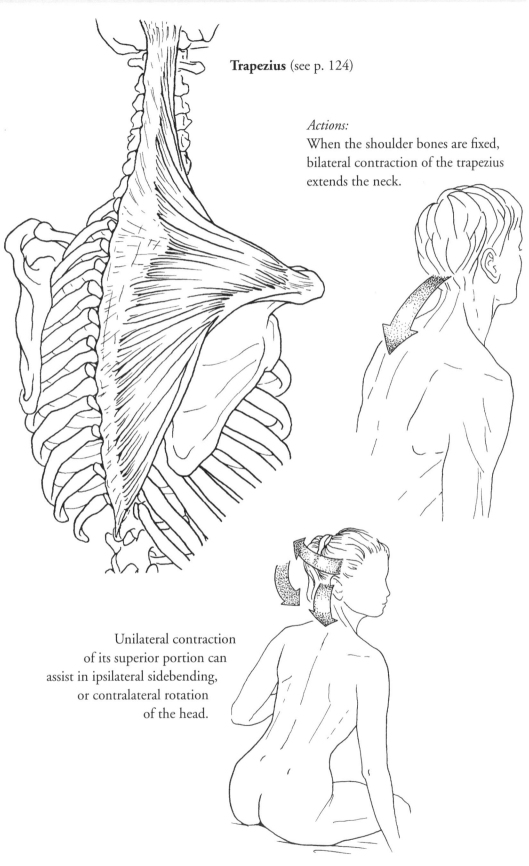

Trapezius (see p. 124)

Actions:
When the shoulder bones are fixed, bilateral contraction of the trapezius extends the neck.

Unilateral contraction
of its superior portion can
assist in ipsilateral sidebending,
or contralateral rotation
of the head.

Anterior and lateral neck muscles

Several deep muscles which run along the cervical spine
can be found on the anterior and lateral sides of the neck.

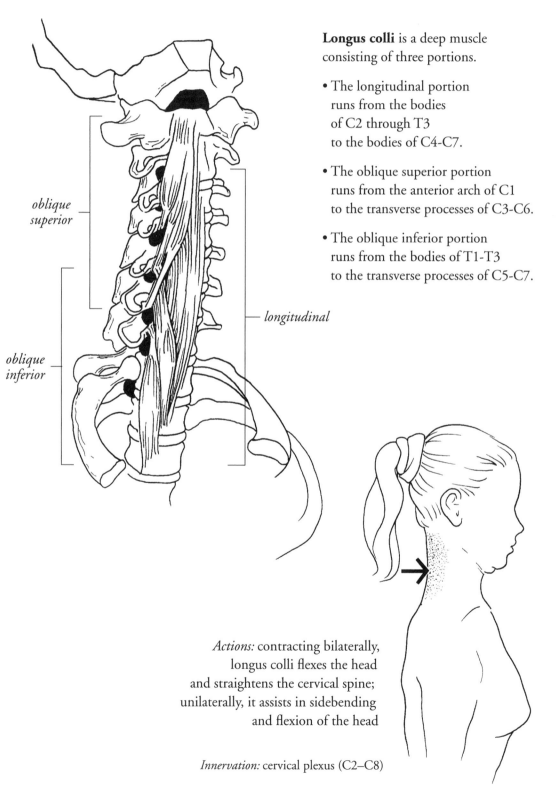

Longus colli is a deep muscle
consisting of three portions.

- The longitudinal portion
  runs from the bodies
  of C2 through T3
  to the bodies of C4-C7.

- The oblique superior portion
  runs from the anterior arch of C1
  to the transverse processes of C3-C6.

- The oblique inferior portion
  runs from the bodies of T1-T3
  to the transverse processes of C5-C7.

*oblique
superior*

*oblique
inferior*

longitudinal

Actions: contracting bilaterally,
longus colli flexes the head
and straightens the cervical spine;
unilaterally, it assists in sidebending
and flexion of the head

Innervation: cervical plexus (C2–C8)

The following muscles attach to the cervical spine and the occiput
(the bone at the posterior base of the skull).

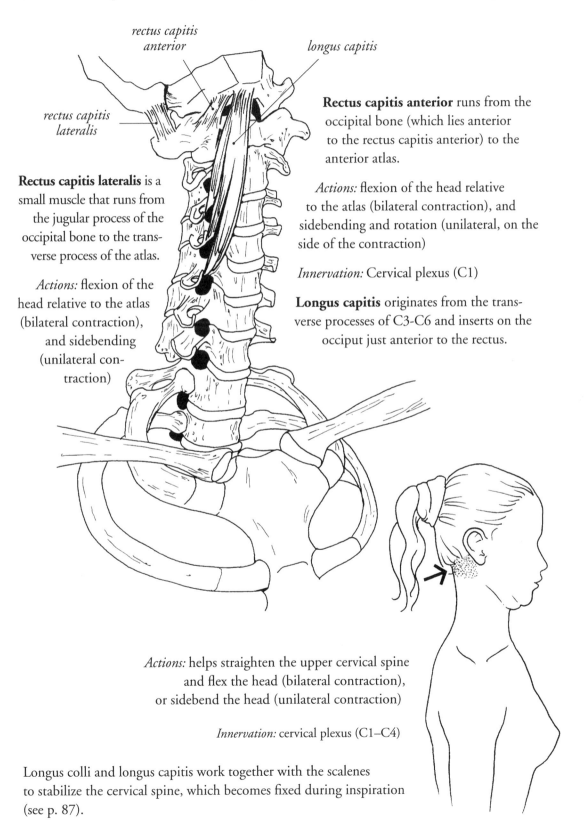

*rectus capitis
anterior*

longus capitis

*rectus capitis
lateralis*

Rectus capitis anterior runs from the
occipital bone (which lies anterior
to the rectus capitis anterior) to the
anterior atlas.

Actions: flexion of the head relative
to the atlas (bilateral contraction), and
sidebending and rotation (unilateral, on the
side of the contraction)

Innervation: Cervical plexus (C1)

Longus capitis originates from the transverse processes of C3-C6 and inserts on the
occiput just anterior to the rectus.

Rectus capitis lateralis is a
small muscle that runs from
the jugular process of the
occipital bone to the transverse process of the atlas.

Actions: flexion of the
head relative to the atlas
(bilateral contraction),
and sidebending
(unilateral con-
traction)

Actions: helps straighten the upper cervical spine
and flex the head (bilateral contraction),
or sidebend the head (unilateral contraction)

Innervation: cervical plexus (C1–C4)

Longus colli and longus capitis work together with the scalenes
to stabilize the cervical spine, which becomes fixed during inspiration
(see p. 87).

The following muscles run from the
cervical vertebrae to the first two ribs.

The **scalenes** consist of three muscles:

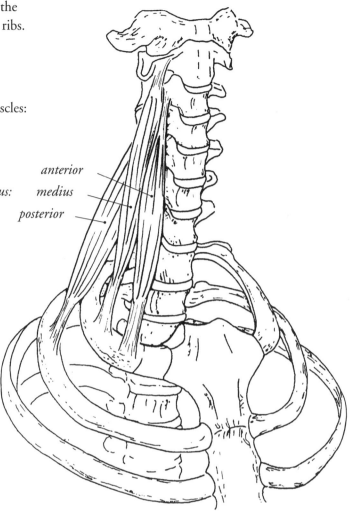

scalenus: anterior
             medius
             posterior

Scalenus anterior and
scalenus medius originate
from the transverse processes of
C3-C6 and C2-C7 respectively,
and insert close together on the
anterior part of rib 1.

Scalenus posterior runs
from the transverse processes
of C4-C6 to the lateral
surface of rib 2.

Innervation: brachial plexus
(C4–C8)

The orientation of scalenus posterior
is almost vertical, whereas scalenus anterior
runs obliquely forward.

Actions: when the ribs are fixed,
   unilateral contraction of the scalenes (especially posterior)
      produces sidebending of the cervical spine;
         medius and anterior also produce some
            contralateral rotation.

Bilateral contraction
of medius and anterior
accentuates the curvature
of the spine.

When the cervical (and upper thoracic) spine is fixed,
contraction of the scalenes elevates ribs 1 and 2,
   assisting in inspiration.

Longus colli and the scalenes play an important role
in stabilizing the cervical spine during this action.

Here, we will simply list all the muscles that pass below
and above the hyoid bone. A detailed study of this bone
and its muscles are beyond the scope of this book.

Suprahyoid muscles:
- hyoglossus
- geniohyoid
- mylohyoid
- digastric
- stylohyoid

Infrahyoid muscles:
- sternohyoid
- thyrohyoid
- omohyoid

Among other actions, these muscles assist
with the flexion of the head above the neck
and thorax.

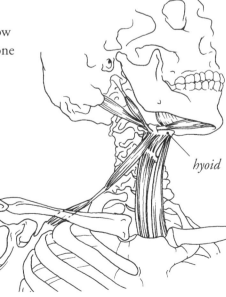

hyoid

Superficial to the muscles mentioned on the previous pages lies another muscle:

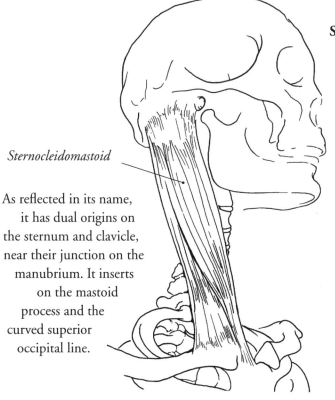

Sternocleidomastoid (SCM)

is the largest and most important anterior neck muscle.

It can be clearly seen externally when the head is turned.

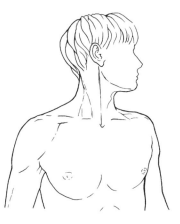

Sternocleidomastoid

As reflected in its name, it has dual origins on the sternum and clavicle, near their junction on the manubrium. It inserts on the mastoid process and the curved superior occipital line.

Innervation: spinal accessory (11th cranial nerve) cervical plexus (C1–C2)

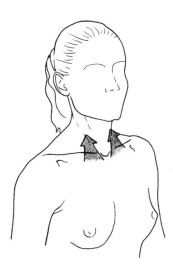

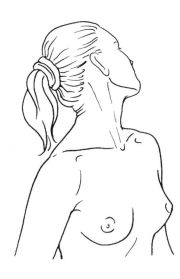

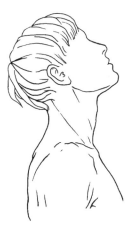

Actions: When the skull is fixed, SCM elevates the sternum and clavicle, and thereby assists in inspiration.

When the thoracic cage is fixed, unilateral contraction of SCM causes ipsilateral sidebending and contralateral rotation of the head, as well as extension.

Bilateral contraction results in extension of the head, accentuating cervical lordosis.

Muscles of the thorax

The **intercostal muscles**
occupy the spaces between adjacent ribs,
and are arranged in two thin layers.

The fibers of the **internal intercostals**
run downward and backward
to the upper border of the rib below.

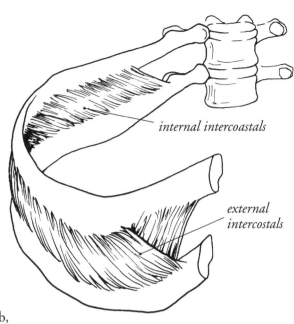

internal intercoastals

*external
intercostals*

The **external intercostals** run downward and *forward* to the rib below.

Actions: The intercostals form a
muscular sheet, which joins the ribs
to each other, making the thoracic cage
a contiguous entity. Thus a muscle, e.g.,
the anterior scalene, pulling on the first rib,
will also pull the entire ribcage, because of the intercostals.

Innervation: 1st–11th intercostal nerves

Levatores costarum run from the
transverse process of a thoracic vertebra
to the tubercle of a rib located
one or two levels below.

Action: assists in rotation of
the spine or elevation of the ribs,
depending on which end is fixed

levatores costarum

Innervation: posterior branches of the spinal nerves

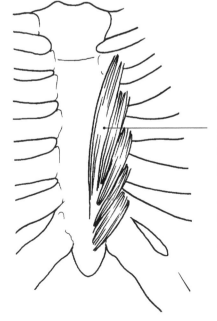

Transversus thoracis originates from the
posterior surface of the lower sternum and
xiphoid process, and runs superolaterally to
insert on the cartilages of ribs 2 through 6.

Action: contraction lowers these ribs,
assisting in expiration

Innervation: 2nd–6th intercostal nerves

Pectoralis major and serratus anterior
are described together with the muscles
of the shoulder (see p. 120, 130).

Diaphragm

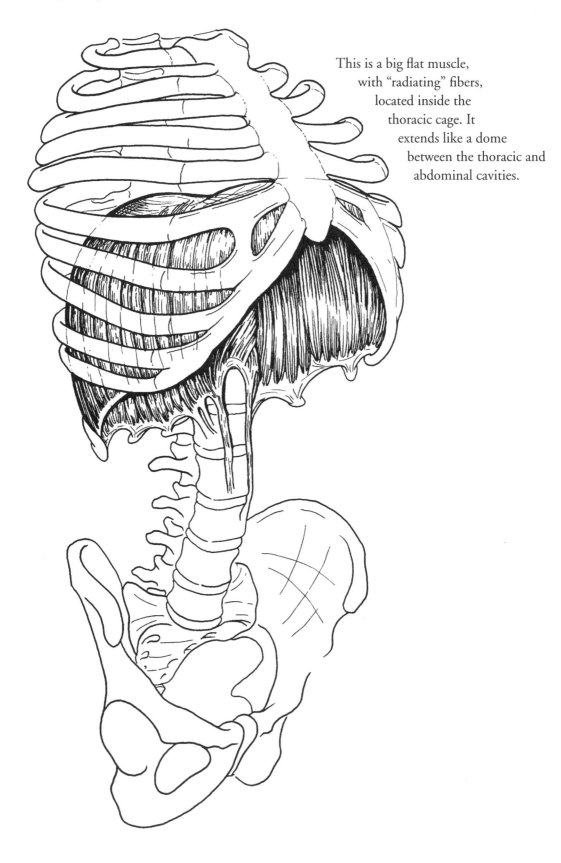

This is a big flat muscle,
with "radiating" fibers,
located inside the
thoracic cage. It
extends like a dome
between the thoracic and
abdominal cavities.

The diaphragm inserts on its own **central tendon**, which is composed of strong interlacing fibrous tissue and shaped somewhat like a three-leafed clover.

central tendon

Openings for:

esophagus

inferior vena cava

aorta

In the middle and near the vertebral column, the diaphragm is pierced by three large openings for the inferior vena cava, esophagus, and descending aorta.

The muscular portion of the diaphragm has three major origins, all inserting on the central tendon:

- The **sternal origin** is from the xiphoid process.

- The **costal origin** is from the deep surfaces of ribs 7-12 and their cartilages; the fibers of attachment interdigitate with those of the transversus abdominis muscle.

- The **vertebral origin** consists of a "right crus" arising directly from the bodies of L1-L3, a "left crus" arising from the bodies of L1-L2, and five **arcuate ligaments.**

  —The single median arcuate ligament joins the two crura at the midline and arches over the abdominal entrance of the aorta.

  —The paired medial arcuate ligaments extend from the body of L1 to its tranverse processes, arching over the psoas major muscle.

  —The paired lateral arcuate ligaments extend from the transverse processes of L1 to rib 12, arching over the quadrates lurnborum muscle.

Action: The diaphragm is the principal muscle of inspiration (see p. 100).

Innervation: phrenic nerves (C3–C5)

Posterior muscles of trunk

Starting from the sides of the lumbar vertebrae,
we find two muscles, psoas major and quadratus lumborum.

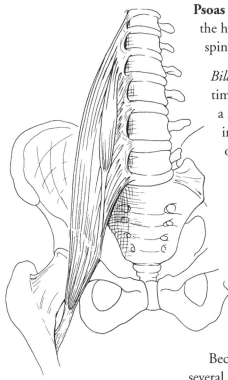

Psoas major is described on page 234 in connection with the hip muscles. Here, we shall consider its action on the spine, i.e., when the thigh is fixed.

Bilateral contraction: For the longest time, the psoas was described as a lumbar muscle involved in increasing lordosis, because of the oblique orientation of its fibers.

But we can see that this multi-articular muscle (which passes over eight joints, six of which are intervertebral) can have a more complex action on the level of the lumbar spine.

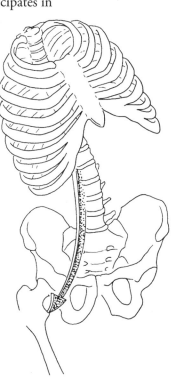

Because of its placement on several levels in the convex area of the lumbar spine, this muscle participates in straightening the spine, in combination with the posterior transversospinalis muscles. By contracting together, these four bundles can act to erect (straighten) the lumbar spine, rather than increasing lordosis. This was discovered in electromyographic recordings taken from moving subjects.

Unilateral contraction: The psoas muscle's unilateral action consists of pulling the lumbar spine into sidebending, flexion, and rotation of the side opposite the contraction.

Innervation: lumbar plexus (L1-L3)

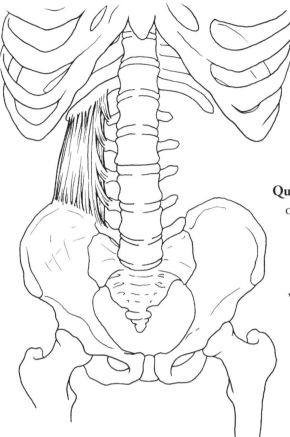

Quadratus lumborum
originates from the posterior iliac crest, and inserts on rib 12 and the transverse processes of L1-L5.

It is composed of both vertical and oblique fibers.

Actions: When the pelvis is fixed, contraction of this muscle pulls on rib 12 and the other ribs along with it. This causes sidebending of the lumbar spine and ribcage.

It is also an expiratory muscle.

When the ribs and spine are fixed, it raises the pelvis on one side.

Innervation: lumbar plexus (T12/L1-L3)

Abdominal muscles

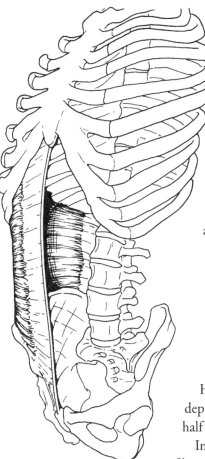

There are four paired muscles in the abdominal wall. They fill all the space (front, sides, and back) between the ribcage and pelvis.

Transversus abdominis is the deepest of the four. It attaches below to the **inguinal ligament** and the iliac crest; posteriorly to the five lumbar vertebrae; above to the inner surfaces of the last seven ribs (where it interdigitates with fibers of the diaphragm); and anteriorly to the **linea alba.**

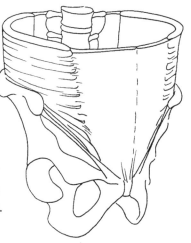

Here is a schematic depiction of the lower half of the transversus. In this depiction, the fibers of the transversus are essentially horizontal. They terminate anteriorly in a broad aponeurosis.

Actions: Contraction of these circular fibers reduces the diameter of the abdomen. When the vertebrae are fixed, it pulls in the belly. When the anterior aponeurosis is fixed, it increases lordosis of the lumbar spine.

An easy way to feel the action of the transversus is to wrap your hands around the sides of your abdomen and cough.

Innervation: intercostal nerves (T7–T12), ilioinguinal and iliohypogastric nerves (L1)

The **internal oblique** lies between the transversus and external oblique. It is attached below to the inguinal ligament and the iliac crest; posteriorly to the lumbodorsal fascia; above, to the lower four ribs; and anteriorly to a very broad aponeurosis.

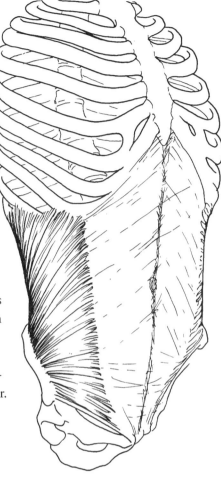

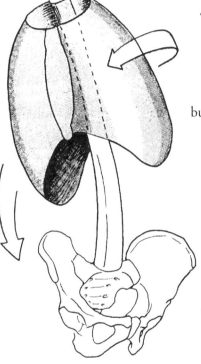

The fibers run in various directions depending on their location, but the "average" direction is anterosuperior.

Actions: Unilateral contraction of the internal oblique results in sidebending or ipsilateral rotation of the spine and ribcage. If the pelvis is fixed, it acts on the ribs, and vice versa.

When the pelvis is fixed, bilateral contraction causes compression of the abdomen and assists in flexion of the trunk.

When the vertebrae and the pelvis are fixed, it lowers the ribs and moves them backward. This is how the muscle is involved in expiration (not shown here).

Innervation:
intercostal nerves (T9-T12),
ilioinguinal and iliohypogastric nerves (L1)

The **external oblique** is attached above to the outer surfaces of ribs 5-12 (where its fibers intertwine with those of the serratus anterior) and to the ilioinguinal ligament. In front and below, it forms a broad aponeurosis ending at (and contributing to) the linea alba and inguinal ligament.

The "average" direction of the fibers is anteroinferior, i.e., perpendicular to those of the internal oblique.

Actions:
Unilateral contraction of the external oblique results in side-bending and contralateral rotation of the spine and ribcage.

Bilateral contraction causes flexion of the trunk. The oblique muscles work synergetically in rotation of the trunk.

For instance, rotation of the trunk to the left (combined with flexion) involves simultaneous contraction of the left internal oblique and right external oblique.

When the pelvis is fixed, it lowers the ribs; it is then an expiratory muscle (not shown here).

Innervation: intercostal nerves (T5-T12), ilioinguinal and iliohypogastric nerves (L1)

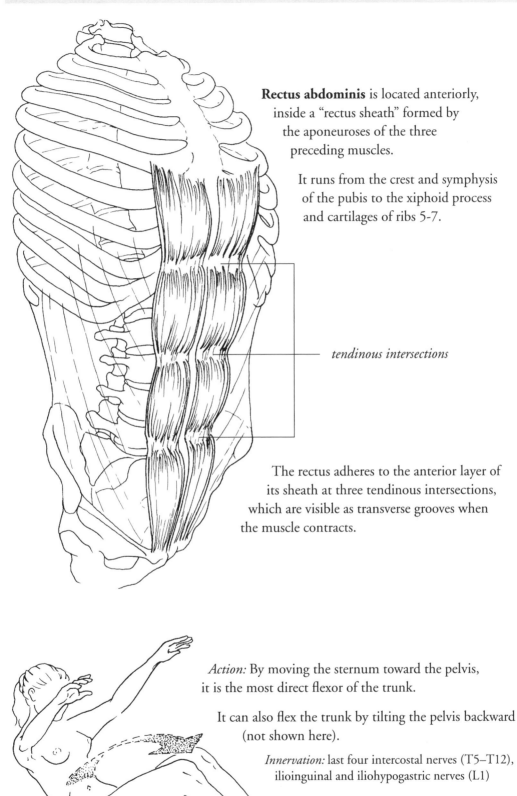

Rectus abdominis is located anteriorly, inside a "rectus sheath" formed by the aponeuroses of the three preceding muscles.

It runs from the crest and symphysis of the pubis to the xiphoid process and cartilages of ribs 5-7.

tendinous intersections

The rectus adheres to the anterior layer of its sheath at three tendinous intersections, which are visible as transverse grooves when the muscle contracts.

Action: By moving the sternum toward the pelvis, it is the most direct flexor of the trunk.

It can also flex the trunk by tilting the pelvis backward (not shown here).

Innervation: last four intercostal nerves (T5–T12), ilioinguinal and iliohypogastric nerves (L1)

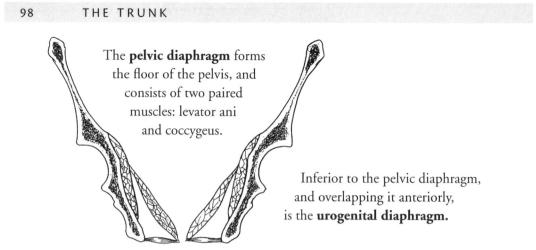

The **pelvic diaphragm** forms the floor of the pelvis, and consists of two paired muscles: levator ani and coccygeus.

Inferior to the pelvic diaphragm, and overlapping it anteriorly, is the **urogenital diaphragm.**

Levator ani originates from fascia covering the obturator internus, along a line extending from the posterior body of the pubis to the ischial spine. The fibers from the left and right sides of this muscle run inferomedially and meet each other at the midline. Some posterior fibers insert on the coccyx and lower sacrum.

Levator ani contains an opening for the anal canal.

The anterior portion of this muscle is different in the female pelvis, where it has an opening to accommodate the vulva (as shown in the illustration). In the male pelvis, this area is closed (not shown here).

Innervation: accessory branch of sacral plexus (S3)

coccygeus muscle

anal canal

levator ani muscle

opening for vulva

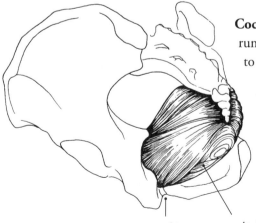

pubic symphysis   *urogenital diaphragm*

Coccygeus is posterior to levator ani, running from the ischial spine to the coccyx and lower sacrum.

Actions: Both of these muscles support the weight of the pelvic organs, and are involved in continence. They also rotate the sacrum backward.

Innervation: accessory branch of pudendal plexus (S4)

Note: These muscles play no role in the positioning of the pelvis on top of the legs, since they do not insert into the femurs.

Abdominal cavity

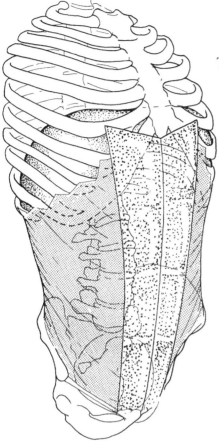

The abdominal cavity is inferior to the thorax and contains the abdominal organs.

It is bounded:

• above by the diaphragm, lower ribs, costal cartilages and sternum

• posteriorly by the lumbar vertebrae

• laterally and anteriorly by the abdominal muscles

• inferiorly by the pelvis and the pelvic and urogenital diaphragm.

Diaphragm and abdominal muscles in respiration

The two large cavities of the trunk, thorax and abdomen, function differently.

The abdomen can be compared to a liquid-filled, flexible container which can change its shape but not its volume (i.e., is non-compressible).

By contrast, the thorax can be compared to a gas-filled container which can change its shape and is compressible.

The diaphragm is like a plunger that moves between these two cavities. In cooperation with the abdominal muscles, it helps control the shape, volume, and pressure of both cavities during many activities (breathing, speaking, coughing, defecating, childbirth, hiccups, etc.)

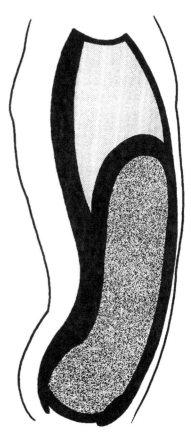

How the diaphragm and abdominal muscles are involved in breathing

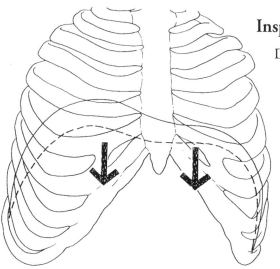

Inspiration

During normal inspiration, contraction of the diaphragm lowers its dome and thereby increases the volume of the thorax and lungs.

The increase in volume results in a decrease of pressure inside the lungs, which causes air to enter from the outside environment.

On the other hand, contraction of the abdominal muscles can cause the abdomen to maintain its shape and resist lowering of the diaphragm.

In this case, the central tendon becomes the fixed point and contraction of the fibers of the diaphragm results in elevation of the ribs, because of

(a) the superomedial orientation of the fibers, and

(b) lateral pressure from the contents of the abdomen, which are being compressed from above.

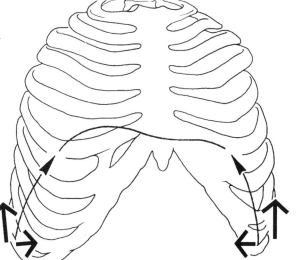

Expiration

During normal expiration, the diaphragm simply relaxes (i.e., the dome moves upward) and the elastic lung tissue returns to its normal size after being stretched during inspiration. The decrease in lung volume results in an increase in pressure, and air is expelled through the nose or mouth. The lungs, however, do not empty completely.

In "forced" (or active) expiration, the expiratory muscles, especially the abdominal muscles, contract and press the abdominal organs upward against the thorax and lower ribcage, further decreasing lung volume and increasing the pressure which drives air out.

No matter how hard "forced" expiration is done, some air will always remain in the lungs. This is called residual volume.

CHAPTER THREE

The Shoulder

. . . .

The shoulder is the area where the arm is attached to the thorax. In contrast to the hip, it involves more than one joint. This complex structure has two important functions:

- It must be very flexible, to allow the hand and arm the huge range of motion which they require.

- It must provide a strong, stable fixed point for certain actions (lifting a heavy object, pushing against resistance, etc.)

The primary joint of the shoulder is the **glenohumeral joint** between the head of the humerus and glenoid cavity of the scapula. However, the scapula itself is an extremely mobile bone, and is connected to the axial skeleton (i.e., sternum) only by the long, thin clavicle. Thus, two other joints are involved in movements of the shoulder: the **acromioclavicular joint** between the distal clavicle and acromion process of the scapula, and the **sternoclavicular joint** between the medial clavicle and manubrium of the sternum.

The shoulder thus comprises three joints, plus important gliding planes. We can distinguish functional areas: the **scapulothoracic** and the **scapulohumeral**.

Landmarks

Some visible and palpable landmarks of the shoulder are shown below.

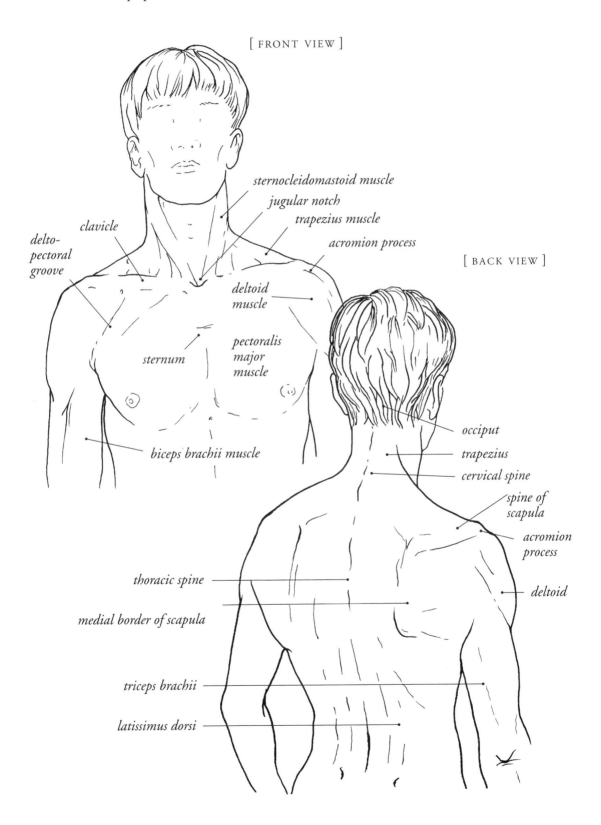

[FRONT VIEW]

sternocleidomastoid muscle

jugular notch

trapezius muscle

acromion process

[BACK VIEW]

clavicle

delto-
pectoral
groove

deltoid
muscle

pectoralis
major
muscle

sternum

occiput

trapezius

cervical spine

spine of
scapula

acromion
process

deltoid

biceps brachii muscle

thoracic spine

medial border of scapula

triceps brachii

latissimus dorsi

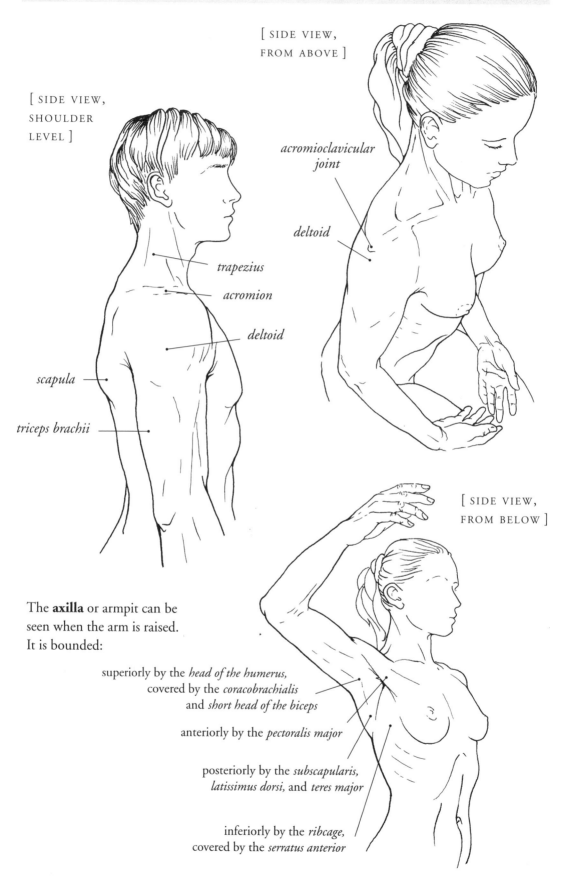

[SIDE VIEW,
FROM ABOVE]

[SIDE VIEW,
SHOULDER
LEVEL]

*acromioclavicular
joint*

deltoid

trapezius

acromion

deltoid

scapula

triceps brachii

[SIDE VIEW,
FROM BELOW]

The **axilla** or armpit can be
seen when the arm is raised.
It is bounded:

superiorly by the *head of the humerus,*
covered by the *coracobrachialis*
and *short head of the biceps*

anteriorly by the *pectoralis major*

posteriorly by the *subscapularis,*
latissimus dorsi, and *teres major*

inferiorly by the *ribcage,*
covered by the *serratus anterior*

Global movements of shoulder

There are two types of movement, because the two functional areas either work individually or together to make them possible. First, we can observe how the shoulder moves on the thorax. These movements make the shoulder:

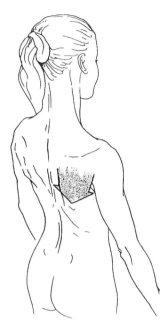

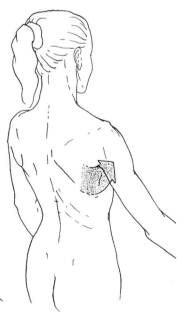

go up: **elevation**

down: **depression**

move away from the spine
(this movement also brings
the shoulder forward): **abduction**

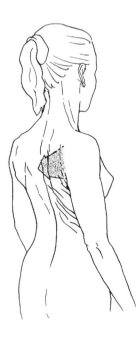

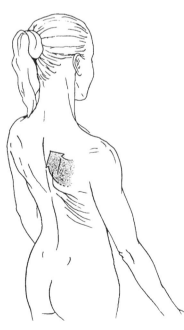

move closer to the spine:
adduction

tilt by moving the inferior angle
of the scapula toward midline:
medial rotation

tilt by moving the inferior
angle away from midline:
lateral rotation.

Second, we can see how the combined actions of the scapula and arm allow the arm to move in many ways:

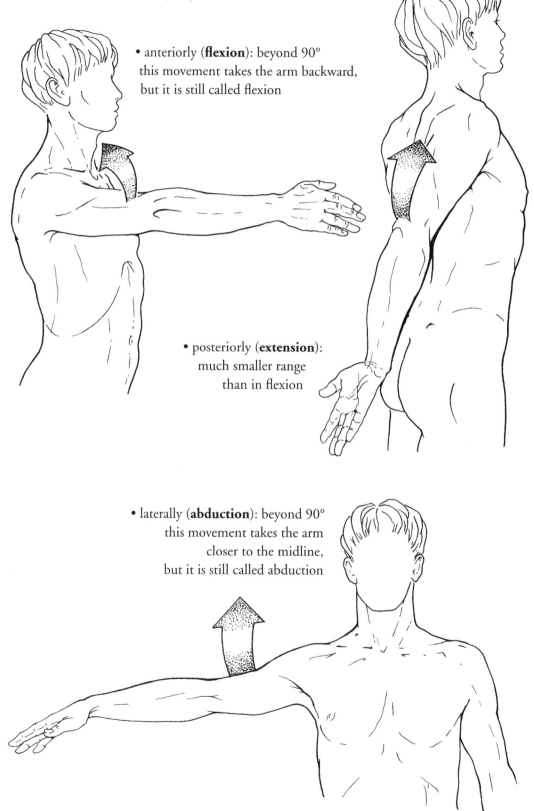

• anteriorly (**flexion**): beyond 90°
this movement takes the arm backward,
but it is still called flexion

• posteriorly (**extension**):
much smaller range
than in flexion

• laterally (**abduction**): beyond 90°
this movement takes the arm
closer to the midline,
but it is still called abduction

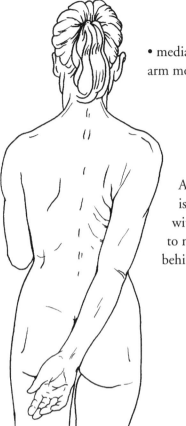

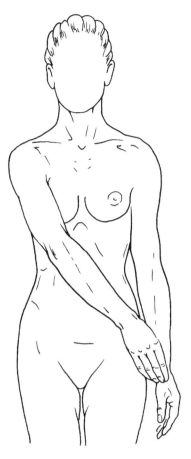

• medially (**adduction**):
arm moves closer to body

Adduction
is combined
with extension
to move the arm
behind the body…

…or with flexion
to move the arm
in front of the body.

Rotation of the humerus
(on its axis) is best visualized
with the elbow bent, to avoid confusion
with pronation and supination
of the forearm (see p. 149):

• **medial rotation**

• **lateral rotation**

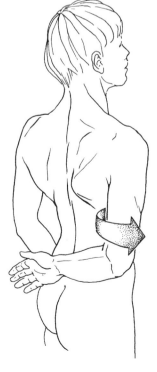

At their extremes,
arm movements affect
the thoracic spine
and ribcage.

Flexion of the arms is associated
with extension of the spine
and "opening" of the
anterior ribcage
(i.e., ribs move apart
from each other).

Extension of the arms
causes flexion of the spine
and "closing" of the ribcage.

Adduction is associated with
ipsilateral sidebending and
closing of the ipsilateral
hemithorax...

...while abduction causes
contralateral sidebending
and opening of the
ipsilateral hemithorax.

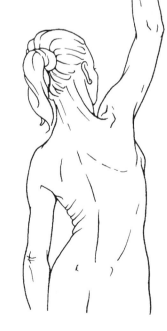

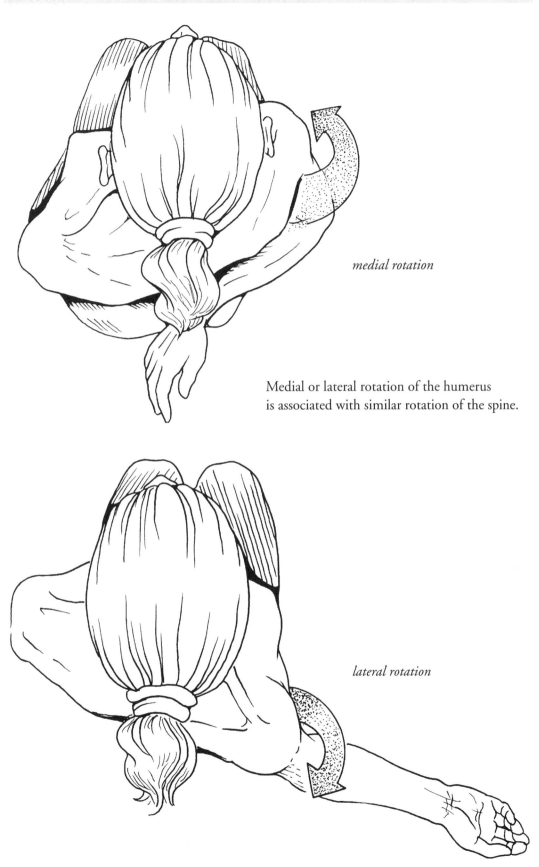

medial rotation

Medial or lateral rotation of the humerus
is associated with similar rotation of the spine.

lateral rotation

Shoulder girdle

The shoulder girdle consists of the scapulae (posteriorly) and the clavicles and sternum (anteriorly).

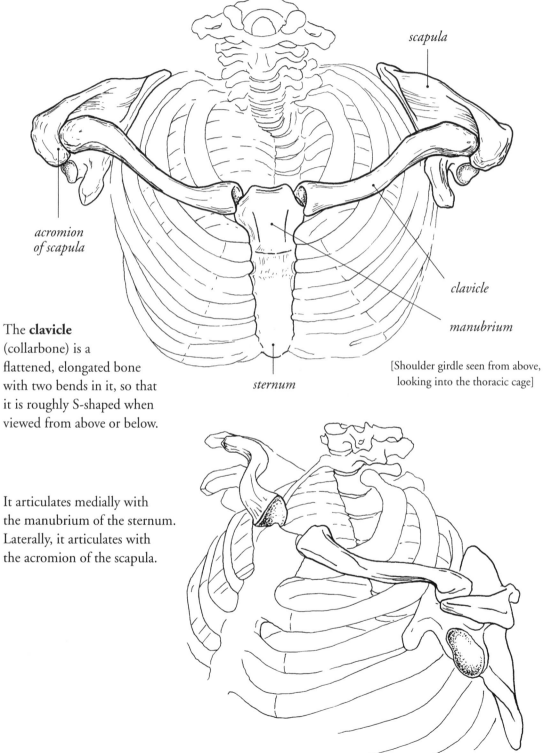

scapula

acromion of scapula

clavicle

manubrium

sternum

[Shoulder girdle seen from above, looking into the thoracic cage]

The **clavicle** (collarbone) is a flattened, elongated bone with two bends in it, so that it is roughly S-shaped when viewed from above or below.

It articulates medially with the manubrium of the sternum. Laterally, it articulates with the acromion of the scapula.

Sternoclavicular joint

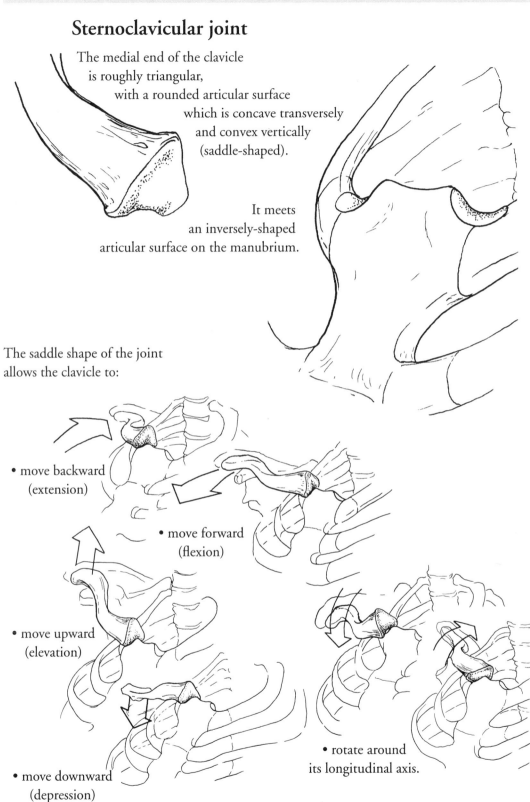

The medial end of the clavicle
is roughly triangular,
with a rounded articular surface
which is concave transversely
and convex vertically
(saddle-shaped).

It meets
an inversely-shaped
articular surface on the manubrium.

The saddle shape of the joint
allows the clavicle to:

- move backward
  (extension)

- move forward
  (flexion)

- move upward
  (elevation)

- move downward
  (depression)

- rotate around
  its longitudinal axis.

These movements generally occur secondarily in association with movements of the scapula.
There are anterior and posterior ligaments (not shown).

Scapula

The scapula, or shoulderblade, is a flat triangular bone
with two surfaces (anterior and posterior), three borders, and three angles.

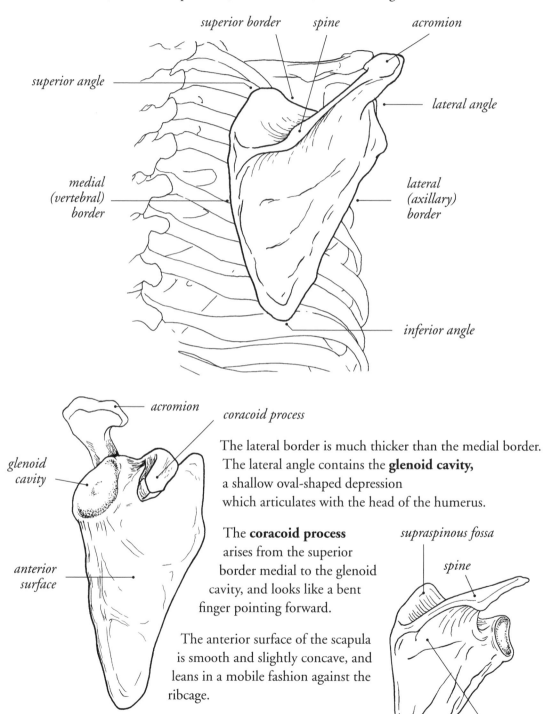

The lateral border is much thicker than the medial border.
The lateral angle contains the **glenoid cavity,**
a shallow oval-shaped depression
which articulates with the head of the humerus.

The **coracoid process**
arises from the superior
border medial to the glenoid
cavity, and looks like a bent
finger pointing forward.

The anterior surface of the scapula
is smooth and slightly concave, and
leans in a mobile fashion against the
ribcage.

The posterior surface is convex. The **spine** is a strong, sharp ridge
running diagonally near the superior border. There are depressions
above and below it called the supraspinous and infraspinous fossae.

The scapular spine is a triangular ridge, which starts roughly perpendicular to the scapular plane.

acromion process

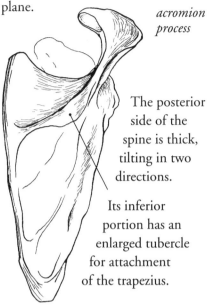

The posterior side of the spine is thick, tilting in two directions.

Its inferior portion has an enlarged tubercle for attachment of the trapezius.

The lateral end of the spine is enlarged and flattened to form the **acromion process.** Note that the acromion is oriented perpendicular to the spine.

The posterior surface of this process can easily be palpated through the skin.

Its anterior surface projects beyond the glenoid cavity; its anterior side consists of an oval articular surface, which articulates with the lateral edge of the clavicle.

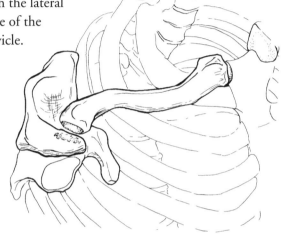

Acromioclavicular joint

The acromion and the lateral end of the clavicle each has a small oval surface where they articulate. Sometimes this joint includes a meniscus (fibrous disk). The shape of the articular surfaces allows some gliding movement, as well as opening and closing of the angle formed by the two bones.

The capsule is loose and the joint is supported by four ligaments:

the superior and inferior acromioclavicular ligaments, which inhibit the opening of the angle between the two bones; and the conoid and trapezoid ligaments, which inhibit the closing of the angle.

These two ligaments are stretched between the coracoid process and the clavicle.

acromioclavicular ligament

conoid and trapezoid ligaments

Movements of shoulder girdle on ribcage

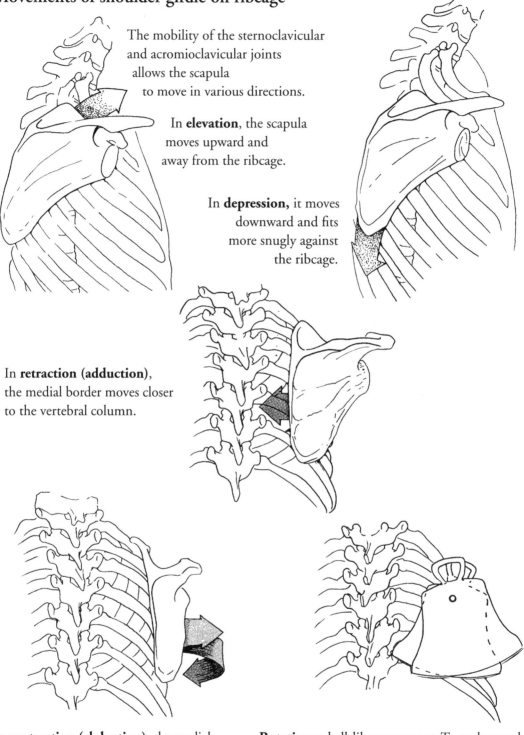

The mobility of the sternoclavicular and acromioclavicular joints allows the scapula to move in various directions.

In **elevation**, the scapula moves upward and away from the ribcage.

In **depression,** it moves downward and fits more snugly against the ribcage.

In **retraction (adduction)**, the medial border moves closer to the vertebral column.

In **protraction (abduction),** the medial border moves away from the vertebral column, and the scapula moves anteriorly at a 45° angle as it glides on the convex thorax.

Rotation: a bell-like movement. To understand this movement, visualize a mobile scapula positioned above the ribcage and moving around an axis, which is vertical to the ribcage and passes under the middle of the scapular spine. The scapula pivots around this axis.

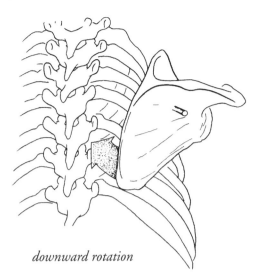

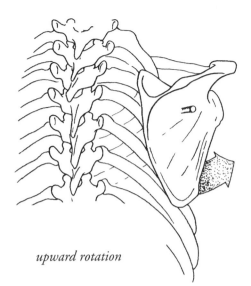

downward rotation

upward rotation

The glenoid cavity of the scapula can point in many directions.
This greatly increases the range of motion of the glenohumeral joint.

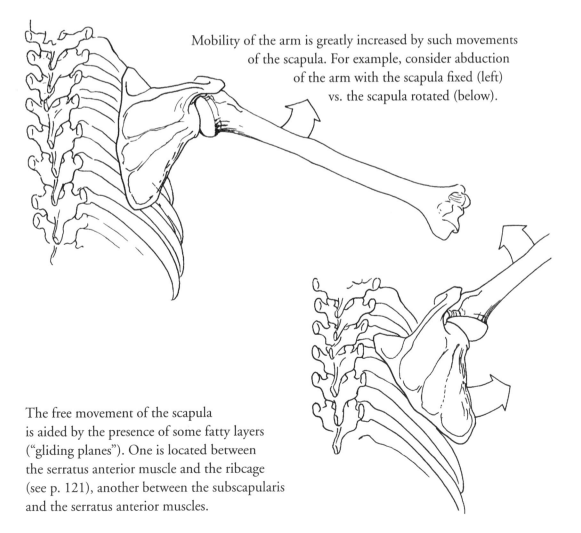

Mobility of the arm is greatly increased by such movements
of the scapula. For example, consider abduction
of the arm with the scapula fixed (left)
vs. the scapula rotated (below).

The free movement of the scapula
is aided by the presence of some fatty layers
("gliding planes"). One is located between
the serratus anterior muscle and the ribcage
(see p. 121), another between the subscapularis
and the serratus anterior muscles.

Humerus

This is the long bone
of the upper arm.

Landmarks of the proximal end
include:

the lateral
greater tubercle

the anterior
lesser tubercle
for muscle attachment

the
bicipital (intertubercular) groove
running between the two tubercles

the medial **head**
(whose spheroid surface
articulates with
the glenoid cavity
of the scapula)

the **anatomical neck**
just below
the head.

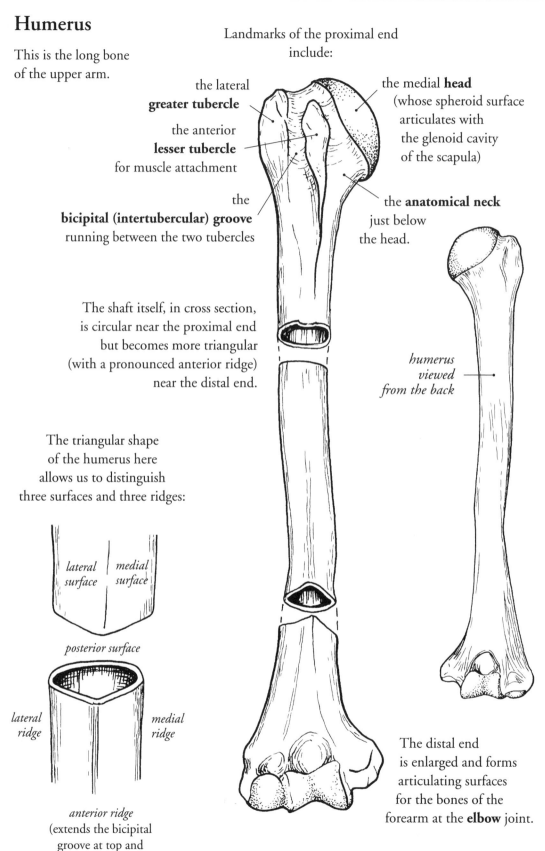

*humerus
viewed
from the back*

The shaft itself, in cross section,
is circular near the proximal end
but becomes more triangular
(with a pronounced anterior ridge)
near the distal end.

The triangular shape
of the humerus here
allows us to distinguish
three surfaces and three ridges:

*lateral
surface* *medial
surface*

posterior surface

*lateral
ridge* *medial
ridge*

anterior ridge
(extends the bicipital
groove at top and
bifurcates at bottom)

The distal end
is enlarged and forms
articulating surfaces
for the bones of the
forearm at the **elbow** joint.

Glenohumeral joint

This is the primary joint of the shoulder, which unites the head of the humerus with the glenoid fossa of the scapula. The articular surfaces consist of:

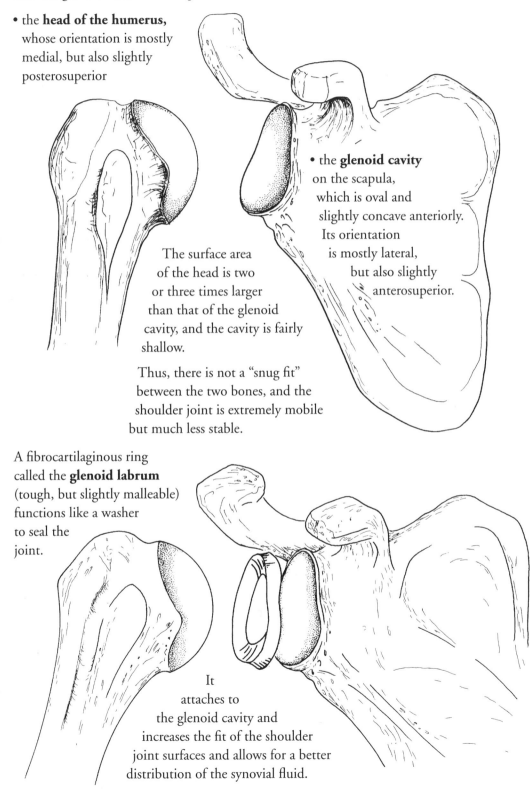

- the **head of the humerus,** whose orientation is mostly medial, but also slightly posterosuperior

- the **glenoid cavity** on the scapula, which is oval and slightly concave anteriorly. Its orientation is mostly lateral, but also slightly anterosuperior.

The surface area of the head is two or three times larger than that of the glenoid cavity, and the cavity is fairly shallow.

Thus, there is not a "snug fit" between the two bones, and the shoulder joint is extremely mobile but much less stable.

A fibrocartilaginous ring called the **glenoid labrum** (tough, but slightly malleable) functions like a washer to seal the joint.

It attaches to the glenoid cavity and increases the fit of the shoulder joint surfaces and allows for a better distribution of the synovial fluid.

The capsule of the glenohumeral joint attaches to the scapula, on the outer rim of the glenoid cavity.

Anterosuperiorly, it goes all the way up to the coracoid process, and encircles the origin of the long head of the biceps at its origin.

The capsule attaches around the head of the humerus. It has many folds, especially in its inferior portion, which gives it a good range of motion for anterior flexion and abduction.

This capsule is reinforced at the top and front by ligaments.

At the top: The **coracohumeral ligament** runs from the coracoid process and forms two bundles, which run to the greater and lesser tubercle of the humerus. This is the strongest ligament of this joint.

In the front:
The **glenohumeral ligaments** run from the border of the glenoid cavity to the anatomical neck of the humerus. They consist of three bundles: superior, middle, and inferior. There are weak areas between these ligaments.

In summary, the capsuloligamentary structure of the shoulder is not very strong. The glenohumeral part of the shoulder is stabilized especially by the deepest muscles, which form a cuff of "active ligaments" around the shoulder, called the rotator cuff (see p. 126-128).

The **coracoacromial ligament** stretches across the scapula and attaches to the acromion and the caracoid process.

This ligament protects the tendon of the supraspinatus.
However, if the humerus is lifted too much, this ligament can rub against the tendon and can, paradoxically, cause it to wear out.

The resting position of the joint (i.e., allowing maximal relaxation of the ligaments) is where the arm is in slight flexion, abduction, and internal rotation.

Shoulder muscles with bony attachments

These bones are grouped into two categories:

- scapulo-thoracic shoulder, which consists of the bones that mobilize the scapula and clavicle with respect to the thorax *(italicized)*

- scapulo-humeral shoulder, which consists of the bones that mobilize the humerus with respect to the scapula.

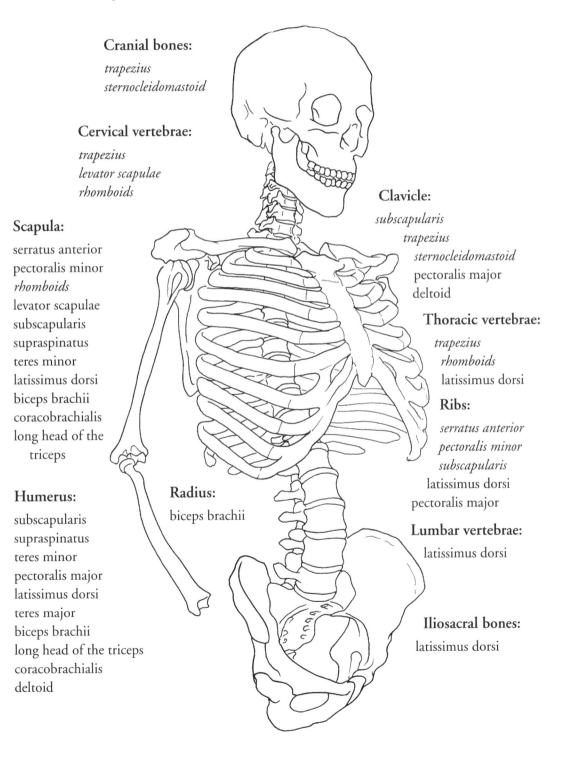

Cranial bones:

trapezius
sternocleidomastoid

Cervical vertebrae:

trapezius
levator scapulae
rhomboids

Scapula:

serratus anterior
pectoralis minor
rhomboids
levator scapulae
subscapularis
supraspinatus
teres minor
latissimus dorsi
biceps brachii
coracobrachialis
long head of the
    triceps

Humerus:

subscapularis
supraspinatus
teres minor
pectoralis major
latissimus dorsi
teres major
biceps brachii
long head of the triceps
coracobrachialis
deltoid

Radius:

biceps brachii

Clavicle:

subscapularis
trapezius
sternocleidomastoid
pectoralis major
deltoid

Thoracic vertebrae:

trapezius
rhomboids
latissimus dorsi

Ribs:

serratus anterior
pectoralis minor
subscapularis
latissimus dorsi
pectoralis major

Lumbar vertebrae:

latissimus dorsi

Iliosacral bones:

latissimus dorsi

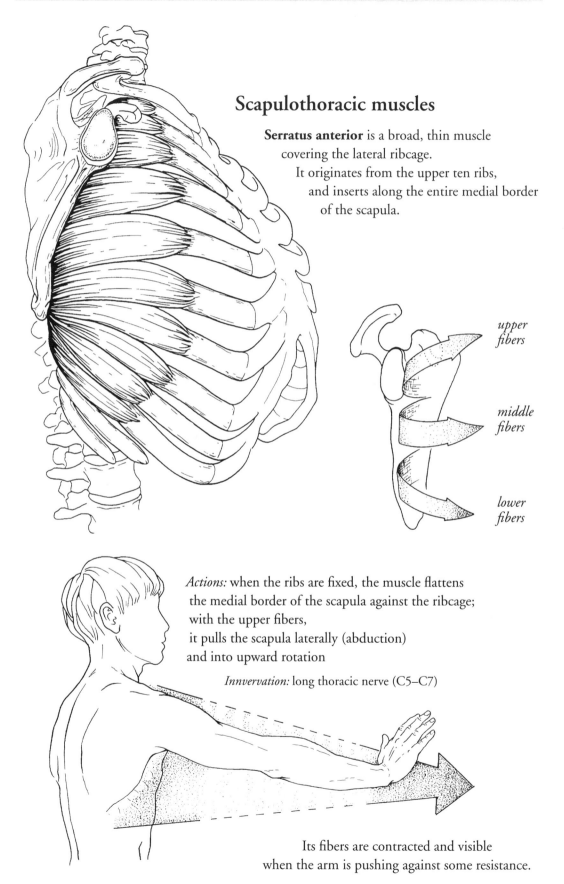

Scapulothoracic muscles

Serratus anterior is a broad, thin muscle covering the lateral ribcage.

It originates from the upper ten ribs, and inserts along the entire medial border of the scapula.

upper fibers

middle fibers

lower fibers

Actions: when the ribs are fixed, the muscle flattens the medial border of the scapula against the ribcage; with the upper fibers, it pulls the scapula laterally (abduction) and into upward rotation

Innervation: long thoracic nerve (C5–C7)

Its fibers are contracted and visible when the arm is pushing against some resistance.

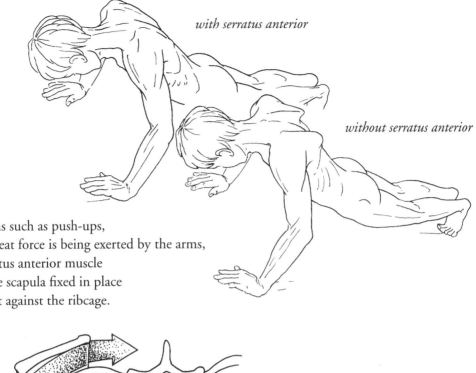

with serratus anterior

without serratus anterior

In actions such as push-ups,
where great force is being exerted by the arms,
the serratus anterior muscle
keeps the scapula fixed in place
and tight against the ribcage.

In such situations, the middle fibers
of trapezius (an adductor)
and serratus (an abductor)
contract simultaneously
to stabilize the scapula.

There are some fatty layers ("gliding planes")
separating serratus from the ribcage
and from the subscapularis muscle.

These increase the mobility of the scapula
and are important in many complex
movements of the shoulder.

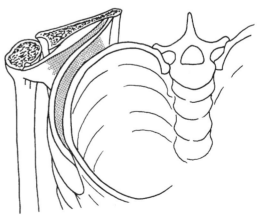

If the scapula is fixed,
the lower fibers of serratus anterior
lift the middle ribs, acting as inspiratory muscles
(not shown here).

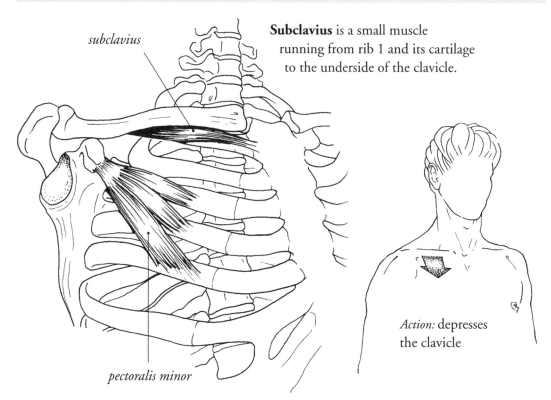

subclavius

Subclavius is a small muscle running from rib 1 and its cartilage to the underside of the clavicle.

pectoralis minor

Action: depresses the clavicle

Pectoralis minor originates from ribs 3-5 and inserts on the coracoid process.

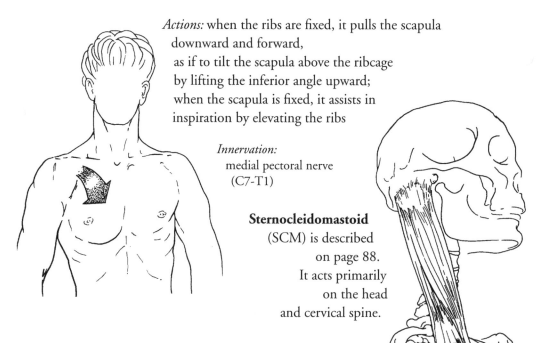

Actions: when the ribs are fixed, it pulls the scapula downward and forward, as if to tilt the scapula above the ribcage by lifting the inferior angle upward; when the scapula is fixed, it assists in inspiration by elevating the ribs

Innervation: medial pectoral nerve (C7-T1)

Sternocleidomastoid (SCM) is described on page 88. It acts primarily on the head and cervical spine.

If the head is fixed, however, SCM can elevate the area where the clavicle and sternum meet, and thereby assist in inspiration.

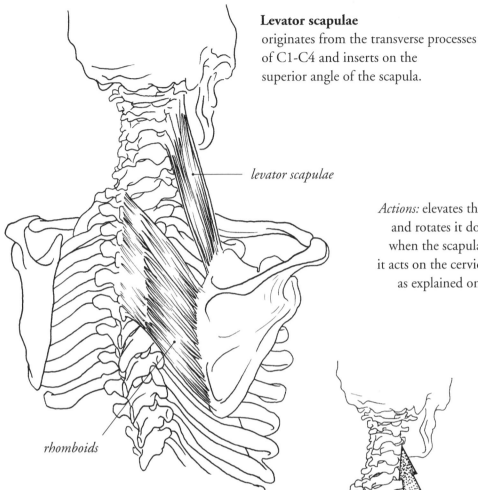

Levator scapulae
originates from the transverse processes
of C1-C4 and inserts on the
superior angle of the scapula.

levator scapulae

Actions: elevates the scapula
and rotates it downward;
when the scapula is fixed,
it acts on the cervical spine,
as explained on page 81

rhomboids

The **rhomboids** (major and minor)
originate from the spinous processes of C7 and T1-T4
and insert on the medial border of the scapula.

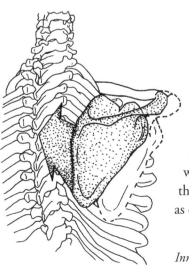

Actions: adduct the scapula
and rotate it downward;
when the scapula is fixed,
they act on the thoracic spine,
as explained on page 82

Innervation: dorsal scapular nerve (C4–C5)

Trapezius is a large, important, diamond-shaped muscle. Its origin is on the occiput, nuchal ligament, and spinous processes of the cervical vertebrae and the thoracic vertebrae down to T12.

Its insertions are on:

- the lateral third of the clavicle and acromion (upper fibers)
- scapular spine (middle fibers)
- a tubercle at the medial end of the scapular spine (lower fibers).

The orientation of these fibers ranges from inferolateral to superolateral.

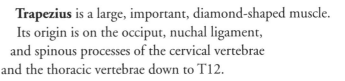

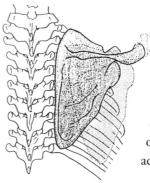

Actions: simultaneous contraction of all the fibers adducts the scapula

The upper fibers by themselves act in elevation and upward rotation of the scapula.

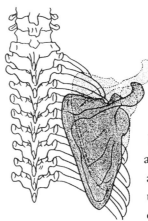

The lower fibers by themselves act in depression and upward rotation of the scapula.

Innervation: spinal accessory nerve and cervical plexus (C2–C4)

The upper fibers are frequently over-solicited in work (e.g., at a computer or typewriter) involving prolonged suspension of the arms. When force needs to be exerted or absorbed by the arm, the middle fibers of trapezius (adductor) act together with serratus anterior (abductor) to stabilize the scapula (see p. 121).

Muscles involved in specific movements of the scapula

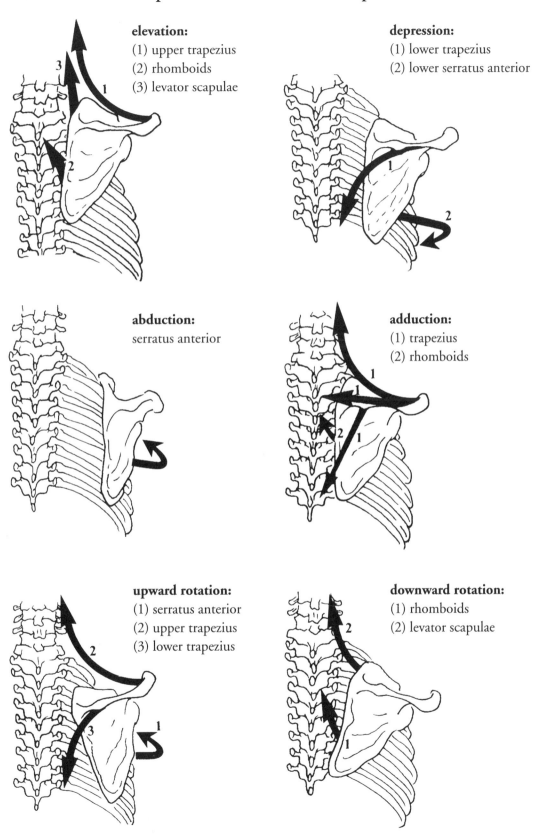

elevation:
(1) upper trapezius
(2) rhomboids
(3) levator scapulae

depression:
(1) lower trapezius
(2) lower serratus anterior

abduction:
serratus anterior

adduction:
(1) trapezius
(2) rhomboids

upward rotation:
(1) serratus anterior
(2) upper trapezius
(3) lower trapezius

downward rotation:
(1) rhomboids
(2) levator scapulae

Deep scapulohumeral muscles of shoulder joint

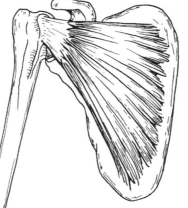

Subscapularis originates from the anterior surface of the scapula and inserts on the lesser tubercle of the humerus. Its fibers converge at the lateral angle of the scapula.

Action: principal muscle of medial rotation of the arm

Innervation: upper subscapular nerve (C5–C6)

Supraspinatus originates from the supraspinous fossa on the posterior scapula. Its tendon passes under the acromioclavicular joint and the ligament which connects the coracoid process to the acromion, and inserts on the highest point on the greater tubercle.

There is a large bursa (closed sac of synovial fluid) surrounding its tendon and separating it from the inferior surface of the acromion and deltoid muscle. This bursa acts as an auxiliary component of the glenohumeral joint. Adhesions here can restrict mobility of the shoulder.

Action: abducts the arm; its action is weak, but is coupled with that of the deltoid (see p. 132)

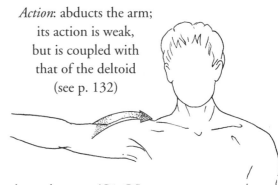

[Viewed from back and above]

Innervation: subscapular nerve (C5–C6)

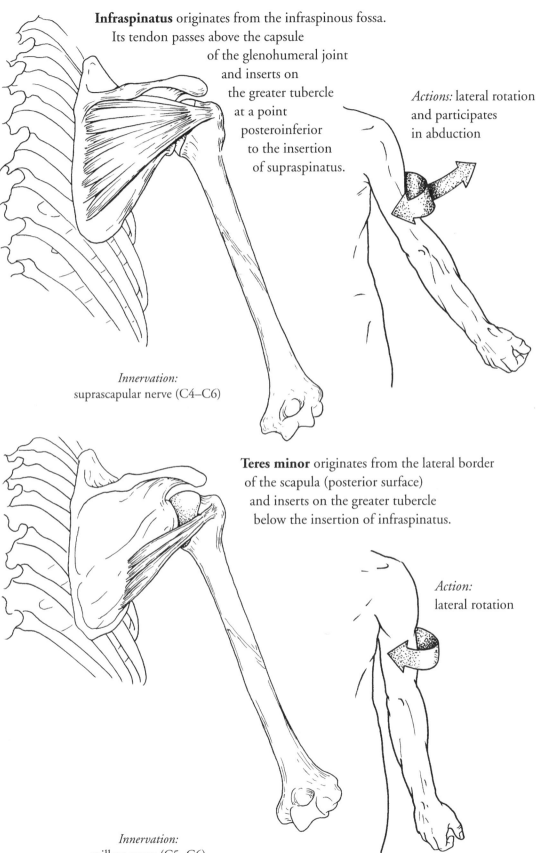

Infraspinatus originates from the infraspinous fossa. Its tendon passes above the capsule of the glenohumeral joint and inserts on the greater tubercle at a point posteroinferior to the insertion of supraspinatus.

Actions: lateral rotation and participates in abduction

Innervation: suprascapular nerve (C4–C6)

Teres minor originates from the lateral border of the scapula (posterior surface) and inserts on the greater tubercle below the insertion of infraspinatus.

Action: lateral rotation

Innervation: axillary nerve (C5–C6)

Rotator cuff muscles

Collectively, these four deep muscles—subscapularis, supraspinatus, infraspinatus, and teres minor—are called the rotator cuff muscles.
Their tendons surround and reinforce the shoulder-joint capsule on three sides.

Apart from their action of mobilizing the humerus, they play an important role as "active ligaments" in providing mobility to the joint.

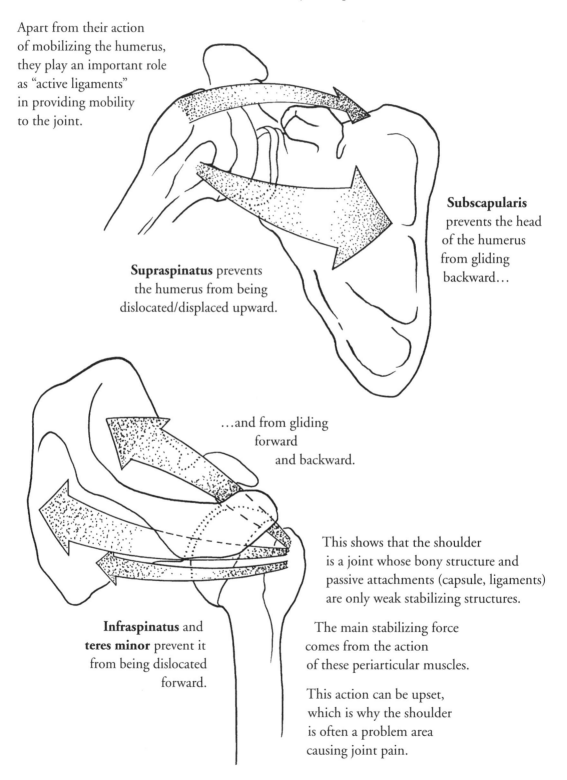

Subscapularis prevents the head of the humerus from gliding backward…

Supraspinatus prevents the humerus from being dislocated/displaced upward.

…and from gliding forward and backward.

This shows that the shoulder is a joint whose bony structure and passive attachments (capsule, ligaments) are only weak stabilizing structures.

Infraspinatus and **teres minor** prevent it from being dislocated forward.

The main stabilizing force comes from the action of these periarticular muscles.

This action can be upset, which is why the shoulder is often a problem area causing joint pain.

Scapulohumeral muscles of shoulder

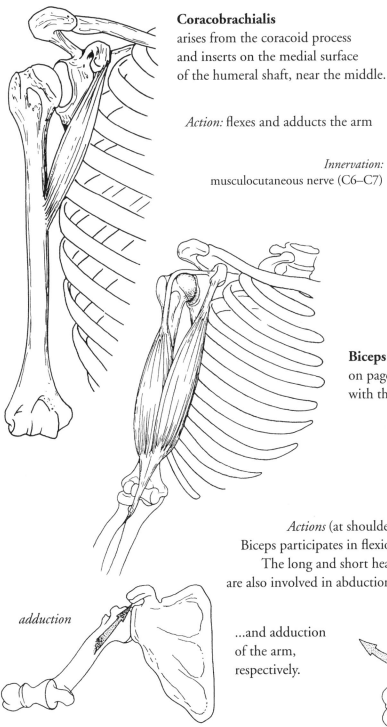

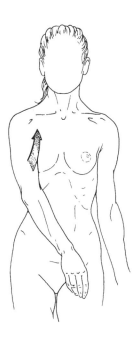

Coracobrachialis
arises from the coracoid process
and inserts on the medial surface
of the humeral shaft, near the middle.

Action: flexes and adducts the arm

Innervation:
musculocutaneous nerve (C6–C7)

Biceps brachii is discussed
on page 147 in connection
with the elbow.

Actions (at shoulder):
Biceps participates in flexion.
The long and short heads
are also involved in abduction...

...and adduction
of the arm,
respectively.

adduction

abduction

Triceps brachii is discussed on page 148
in connection with the elbow.

Action (at shoulder): participates in adduction of the arm

Pectoralis major has a clavicular head from the anterior, medial clavicle, and a sternocostal head from the sternum and costal cartilages 1-6 and rib 7.

The tendon is twisted such that the fibers from the clavicular head insert below those from the sternocostal head on the lateral aspect of the bicipital groove.

When the ribcage is fixed, all the fibers adduct and medially rotate the arm. This is the "hugging" muscle, the muscle of chest suspension. The superior fibers are involved in flexion up to 60°, then the inferior fibers take over and continue flexion up to 0° (see p. 135).

When the shoulder is fixed, the superior fibers lower the clavicle and the inferior fibers participate in inspiration.

When the shoulder is fixed while the arm is flexed, all the fibers are involved in inspiration.

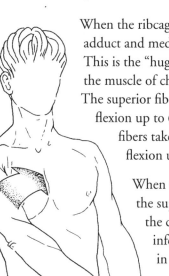

stretching of pectoralis major

Innervation: lateral and medial pectoral nerves (C5–T1)

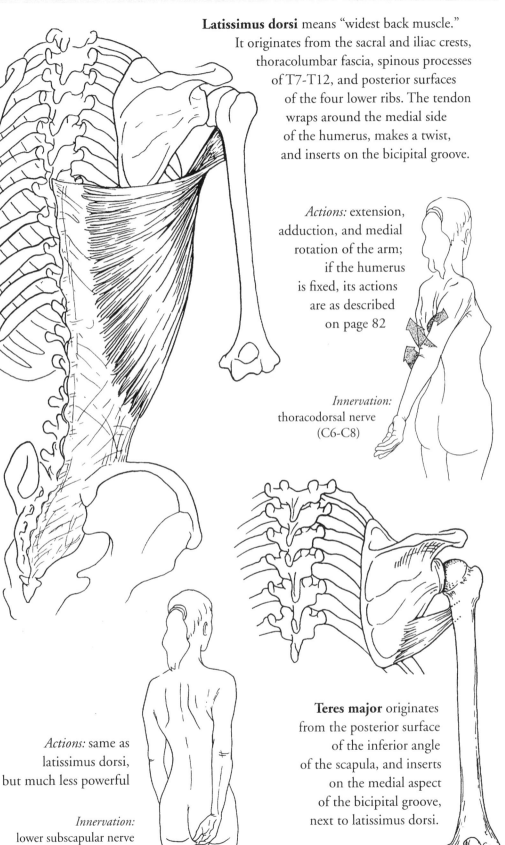

Latissimus dorsi means "widest back muscle." It originates from the sacral and iliac crests, thoracolumbar fascia, spinous processes of T7-T12, and posterior surfaces of the four lower ribs. The tendon wraps around the medial side of the humerus, makes a twist, and inserts on the bicipital groove.

Actions: extension, adduction, and medial rotation of the arm; if the humerus is fixed, its actions are as described on page 82

Innervation: thoracodorsal nerve (C6-C8)

Actions: same as latissimus dorsi, but much less powerful

Innervation: lower subscapular nerve (C6-C7)

Teres major originates from the posterior surface of the inferior angle of the scapula, and inserts on the medial aspect of the bicipital groove, next to latissimus dorsi.

Deltoid is a superficial muscle which gives the shoulder
its characteristic shape. It consists of three groups of fibers:

- The middle fibers attach
  to the lateral border of the acromion.

- The posterior fibers attach
  to the spine of the scapula
  (inferior portion of the posterior border).

- The anterior fibers attach
  to the clavicles
  (lateral third of the anterior scapular border).

These three groups of fibers converge
toward the middle of the arm
and insert on the lateral surface
of the humerus.

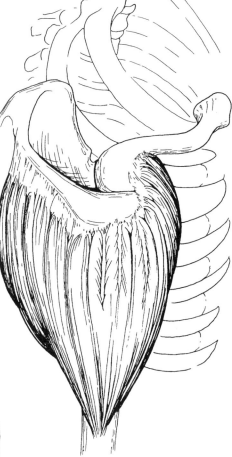

Actions:
contraction of anterior fibers:
flexion and medial rotation
of arm

Innervation: axillary nerve (C5-C6)

contraction of middle fibers:
abduction of arm

contraction of posterior fibers:
extension of arm.

Muscle actions in specific movements of shoulder

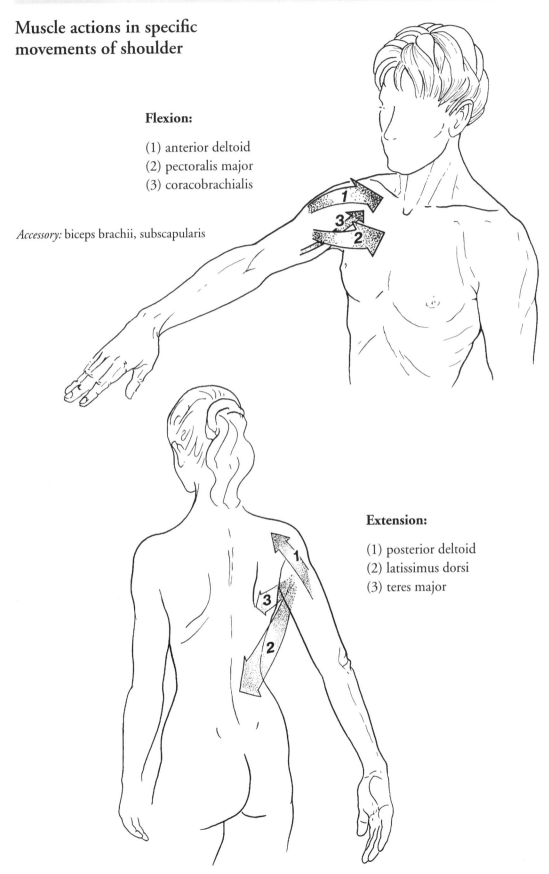

Flexion:

(1) anterior deltoid
(2) pectoralis major
(3) coracobrachialis

Accessory: biceps brachii, subscapularis

Extension:

(1) posterior deltoid
(2) latissimus dorsi
(3) teres major

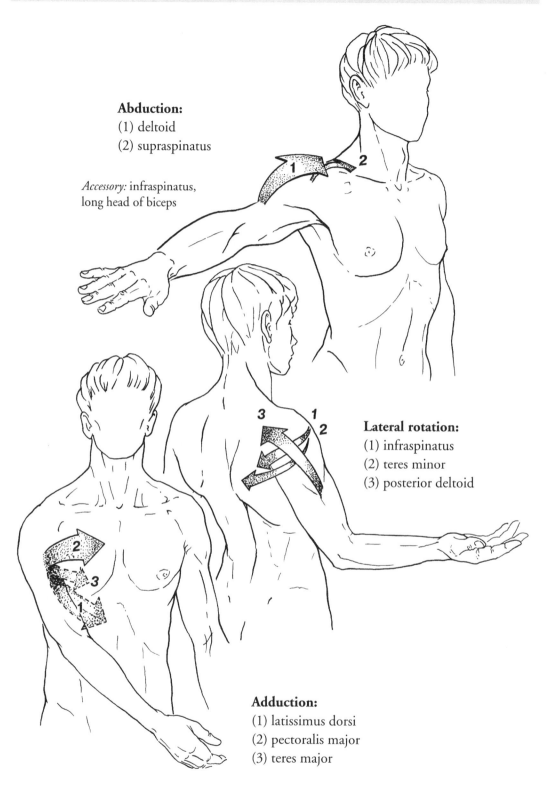

Abduction:
(1) deltoid
(2) supraspinatus

Accessory: infraspinatus,
long head of biceps

Lateral rotation:
(1) infraspinatus
(2) teres minor
(3) posterior deltoid

Adduction:
(1) latissimus dorsi
(2) pectoralis major
(3) teres major

Accessory: teres minor, short head of biceps,
long head of triceps, coracobrachialis

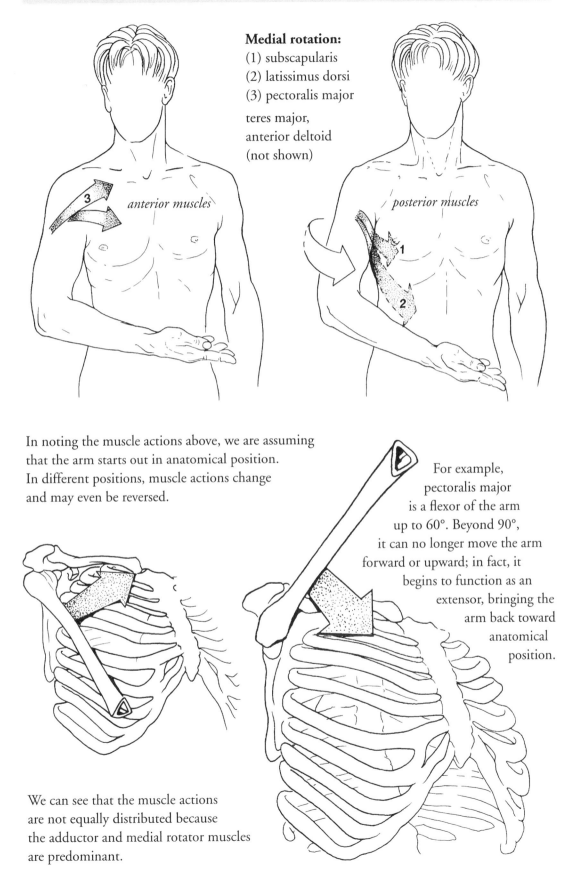

Medial rotation:
(1) subscapularis
(2) latissimus dorsi
(3) pectoralis major

teres major,
anterior deltoid
(not shown)

anterior muscles

posterior muscles

In noting the muscle actions above, we are assuming
that the arm starts out in anatomical position.
In different positions, muscle actions change
and may even be reversed.

For example,
pectoralis major
is a flexor of the arm
up to 60°. Beyond 90°,
it can no longer move the arm
forward or upward; in fact, it
begins to function as an
extensor, bringing the
arm back toward
anatomical
position.

We can see that the muscle actions
are not equally distributed because
the adductor and medial rotator muscles
are predominant.

CHAPTER FOUR

The Elbow

. . . .

The elbow is a joint that serves two functions:

First, it allows the upper limb to fold on itself and to extend, so that the distance between the shoulder and hand can be shortened or lengthened. Through this action, you can, for example, flex your elbow to lift your hand to your mouth, or extend your elbow to reach lower parts of your body or objects located farther away from the shoulder. This is the *flexion-extension* action of the elbow.

Second, the elbow also participates in rotating the forearm around its longitudinal axis, multiplying the possible positions of the hand. This is the *pronation-supination* action of the elbow.

This chapter is accordingly organized into two parts, considering each function in turn.

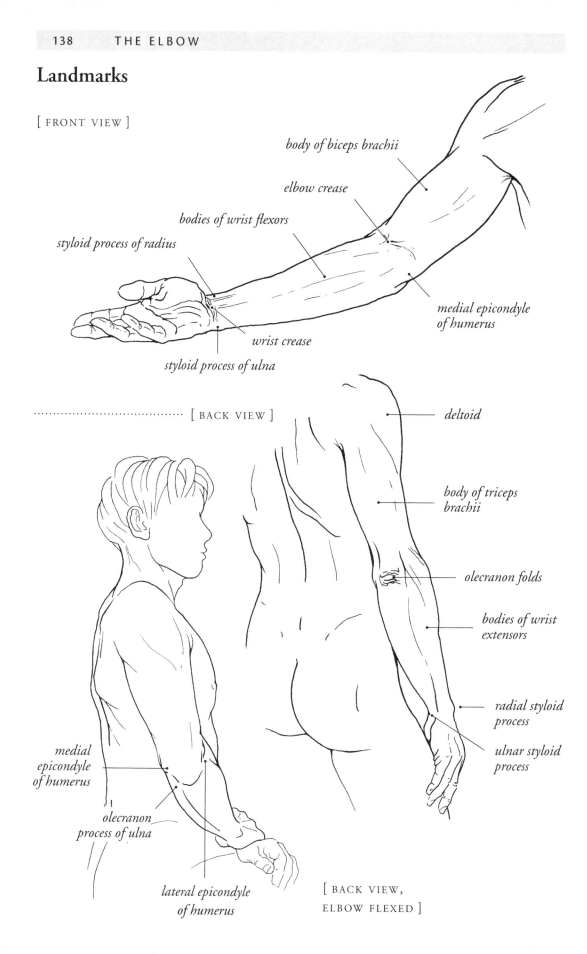

Landmarks

[FRONT VIEW]

body of biceps brachii

elbow crease

bodies of wrist flexors

styloid process of radius

medial epicondyle
of humerus

wrist crease

styloid process of ulna

.... [BACK VIEW]

deltoid

body of triceps
brachii

olecranon folds

bodies of wrist
extensors

radial styloid
process

ulnar styloid
process

medial
epicondyle
of humerus

olecranon
process of ulna

lateral epicondyle
of humerus

[BACK VIEW,
ELBOW FLEXED]

Flexion/extension of the elbow

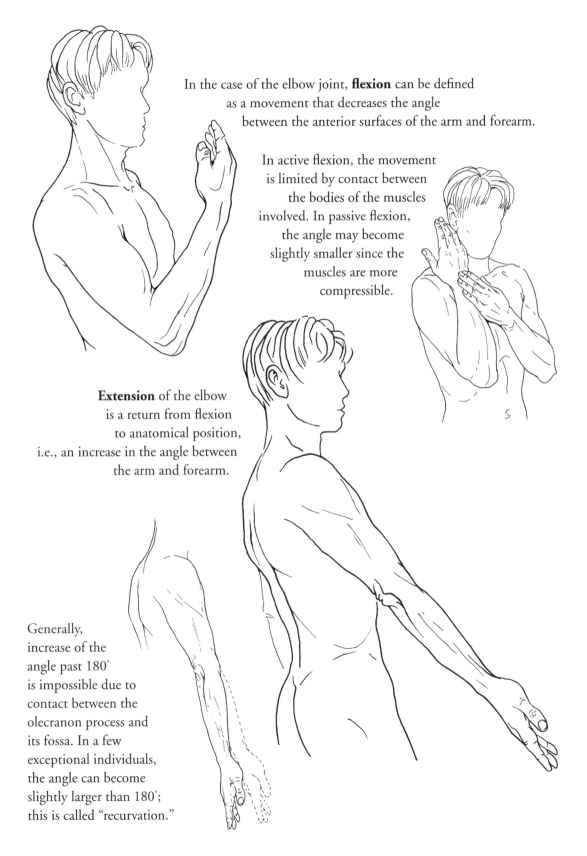

In the case of the elbow joint, **flexion** can be defined as a movement that decreases the angle between the anterior surfaces of the arm and forearm.

In active flexion, the movement is limited by contact between the bodies of the muscles involved. In passive flexion, the angle may become slightly smaller since the muscles are more compressible.

Extension of the elbow is a return from flexion to anatomical position, i.e., an increase in the angle between the arm and forearm.

Generally, increase of the angle past 180° is impossible due to contact between the olecranon process and its fossa. In a few exceptional individuals, the angle can become slightly larger than 180°; this is called "recurvation."

Radius and ulna

RADIUS　　　　　ULNA

The forearm consists of two bones: the radius and the ulna.

These are long bones, each of which has three parts: a shaft and two ends or extremities.

The bones are triangular shaped and therefore have three surfaces and three edges.

The proximal part of the **radius** is small, while its distal part is large.

Its proximal extremity consists of two parts: the **head**, which is covered by cartilage, and the **neck**.

The head consists of a top part (beveled medially) and a circumference.

At its distal part, the medial edge bifurcates, so that the cross section of the bone is quadrangular.

The inferior surface articulates with the wrist.

head

neck

radial
tuberosity

trochlear　olecranon
notch　　　process

coronoid
process

radial
notch

The shaft(s) have cylindrical shapes with three surfaces and three edges:

As　anterior surface
Ps　posterior surface
Ls　lateral surface
　　　(medial on ulna)
Ae　anterior edge
Me　medial edge
　　　(posterior on ulna)
Le　lateral edge

The
proximal
part of the
ulna is large,
while its distal
part is small.
The proximal
part has two large
processes: the
olecranon and the
coronoid process.

As
Ls
Ps

As　　Ms
Ps

Le　　Ae　Me

Le　　Ae　Pe

ulnar
notch

styloid process　articular surface

Its bifurcation has a concave articular surface which articulates with the ulna. This is the **ulnar notch** of the radius.

There is another protuberance at the lateral distal end of the radius: the **styloid process**.

ulnar
head

styloid
process

The distal extremity is called the **ulnar head**. Laterally, it has a **convex articular surface** which articulates with the radius. Medially, it has a protuberance, the **styloid process**.

Attached to the inferior surface is a triangular ligament, which connects the ulna to the bones of the wrist.

Bones and articulating surfaces for flexion and extension of elbow

At the distal end of the humerus the anterior edge bifurcates and the bone flattens out, becomes larger, and curves forward. It has two lateral bony projections, the **medial** and **lateral epicondyles**.

lateral epicondyle

medial epicondyle

coronoid fossa

trochlea

radial fossa

capitulum

Posteriorly, there is an olecranon fossa (not shown).

Hollow areas, where the bone becomes tapered, are common on these articular surfaces. Just above the trochlear eminence is the **coronoid fossa** anteriorly and the **olecranon fossa** posteriorly. Just above the capitulum is the **radial fossa** of the humerus.

These two protuberances delimit a triangular space. At the base of this triangle, there are two articular surfaces: the medial pulley-shaped surface is oblique medioinferiorly and articulates with the trochlear notch of the ulna. This is the **trochlea of the humerus**.

ulna

radius

The lateral rounded surface with a diameter of about 1cm is the **capitulum**.

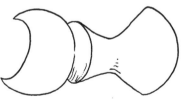

The surfaces between the shafts of the radius and ulna are connected via a thin layer: the **interosseous membrane**.

At the head of the radius, on its upper surface, is a shallow cup or **fovea** which articulates with the capitulum.

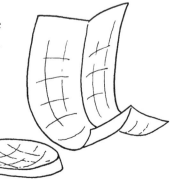

The radius is lined by a beveled layer, which articulates with the interosseous membrane.

Proximal end of the ulna

The **olecranon** consists of five surfaces, including a beak-like projection on its superior surface.

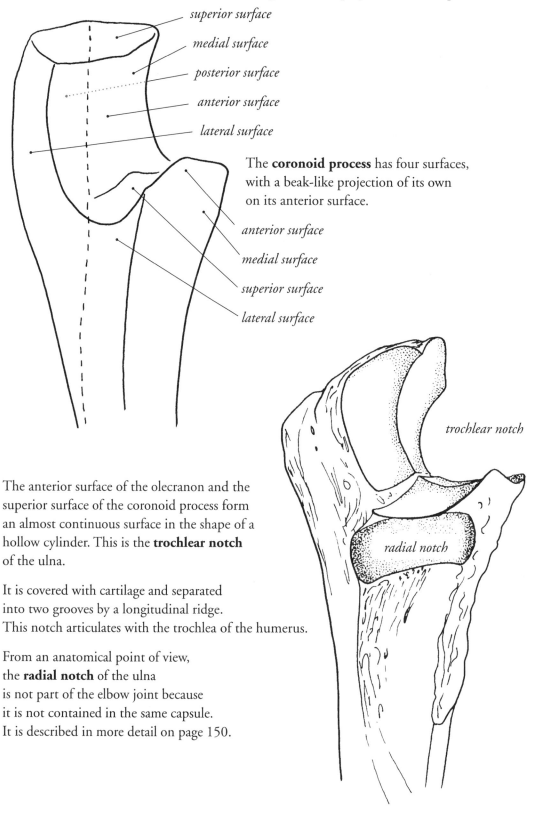

superior surface

medial surface

posterior surface

anterior surface

lateral surface

The **coronoid process** has four surfaces, with a beak-like projection of its own on its anterior surface.

anterior surface

medial surface

superior surface

lateral surface

trochlear notch

radial notch

The anterior surface of the olecranon and the superior surface of the coronoid process form an almost continuous surface in the shape of a hollow cylinder. This is the **trochlear notch** of the ulna.

It is covered with cartilage and separated into two grooves by a longitudinal ridge. This notch articulates with the trochlea of the humerus.

From an anatomical point of view, the **radial notch** of the ulna is not part of the elbow joint because it is not contained in the same capsule. It is described in more detail on page 150.

Joint capsule of elbow

Three bones—the humerus, ulna, and radius—
come together to form the capsule of the elbow joint.

The capsule attaches to the circumference of the coronoid
and olecranon fossae of the humerus and connects
with the medial and lateral epicondyles without
encircling them. It attaches to the circumference
of the neck and the radius. It also attaches to the
circumference of the trochlear notch of the unla.

The capsule is taut in front
(especially laterally), but loose in
back (facilitating flexion).

[The illustration at left shows the joint with bones
spread apart slightly to better view the capsule.]

Ligaments of elbow

The anterior and posterior
elbow ligaments are
minor ligaments:

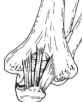

• posteriorly, their fibers
cross over (as shown here
on a flexed elbow).

• anteriorly, they
are fan-shaped and
reinforce the capsule

Thus, they allow
flexion and extension.

The **collateral ligaments**
are the most important
ligaments of the elbow:

The **ulnar
collateral ligament**
arises from the
medial epicondyle
and attaches below to the
medial coronoid process
and olecranon.

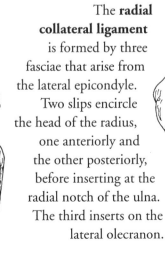

The **radial
collateral ligament**
is formed by three
fasciae that arise from
the lateral epicondyle.
Two slips encircle
the head of the radius,
one anteriorly and
the other posteriorly,
before inserting at the
radial notch of the ulna.
The third inserts on the
lateral olecranon.

Both collateral ligaments prevent any lateral movement of the elbow joint.

Bones for flexion and extension of elbow

The surfaces of the lower part of the humerus
articulate with the ulna and radius. This functional unit
facilitates movement only in the sagittal plane.

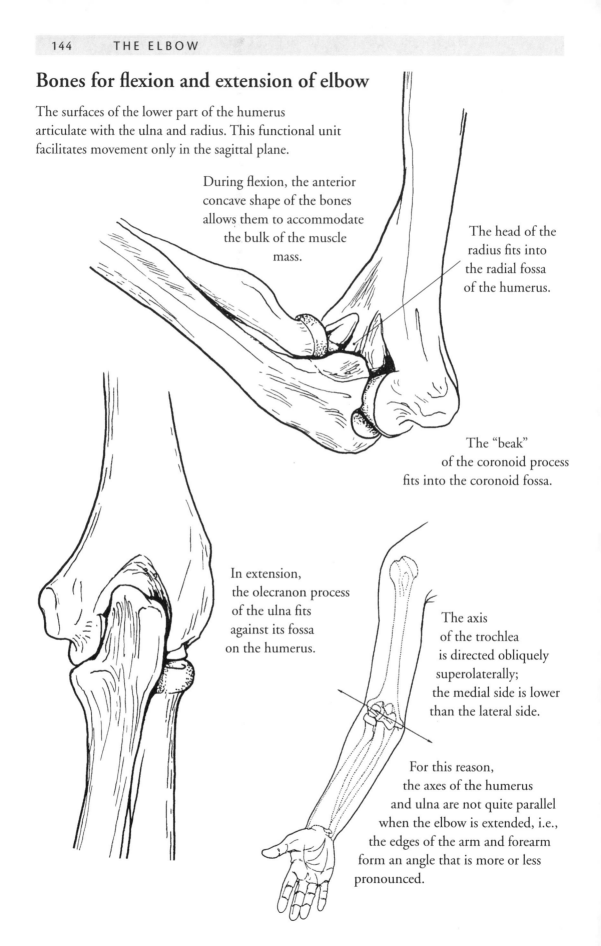

During flexion, the anterior
concave shape of the bones
allows them to accommodate
the bulk of the muscle
mass.

The head of the
radius fits into
the radial fossa
of the humerus.

The "beak"
of the coronoid process
fits into the coronoid fossa.

In extension,
the olecranon process
of the ulna fits
against its fossa
on the humerus.

The axis
of the trochlea
is directed obliquely
superolaterally;
the medial side is lower
than the lateral side.

For this reason,
the axes of the humerus
and ulna are not quite parallel
when the elbow is extended, i.e.,
the edges of the arm and forearm
form an angle that is more or less
pronounced.

Muscles for flexion/extension of elbow with their many bony attachments

Two muscle groups are presented here:
the principal muscles (in **boldface**)
and the secondary muscles,
most of which will be described in Chapter 5 on the wrist and hand.

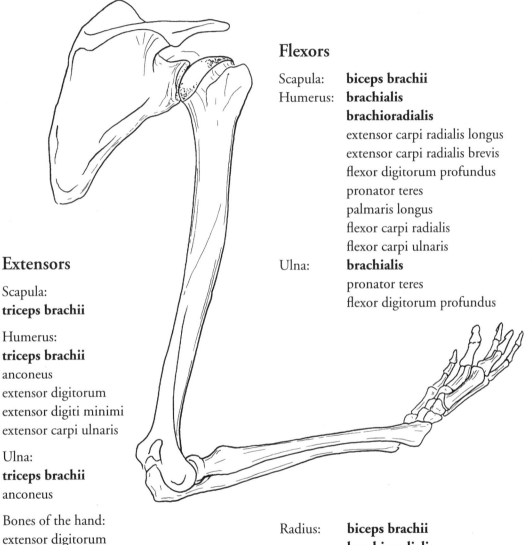

Flexors

Scapula:	**biceps brachii**
Humerus:	**brachialis**
	brachioradialis
	extensor carpi radialis longus
	extensor carpi radialis brevis
	flexor digitorum profundus
	pronator teres
	palmaris longus
	flexor carpi radialis
	flexor carpi ulnaris
Ulna:	**brachialis**
	pronator teres
	flexor digitorum profundus

Extensors

Scapula:
triceps brachii

Humerus:
triceps brachii
anconeus
extensor digitorum
extensor digiti minimi
extensor carpi ulnaris

Ulna:
triceps brachii
anconeus

Bones of the hand:
extensor digitorum
extensor digiti minimi

Radius:	**biceps brachii**
	brachioradialis

Bones of the hand:
extensor carpi radialis longus
extensor carpi radialis brevis
flexor digitorum profundus
palmaris longus
flexor carpi radialis
flexor carpi ulnaris

Muscles for flexion of elbow

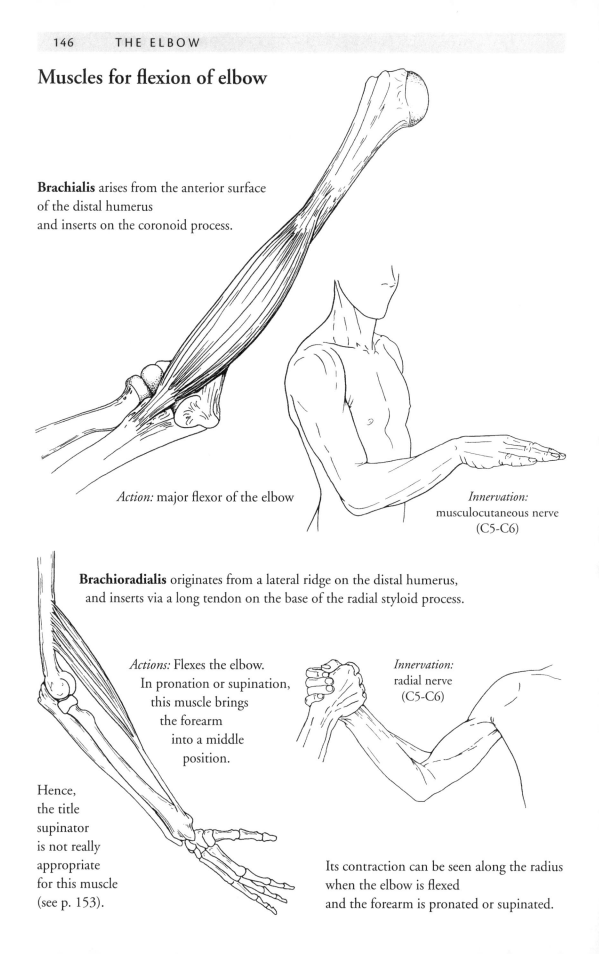

Brachialis arises from the anterior surface
of the distal humerus
and inserts on the coronoid process.

Action: major flexor of the elbow

Innervation:
musculocutaneous nerve
(C5-C6)

Brachioradialis originates from a lateral ridge on the distal humerus,
and inserts via a long tendon on the base of the radial styloid process.

Actions: Flexes the elbow.
In pronation or supination,
this muscle brings
the forearm
into a middle
position.

Innervation:
radial nerve
(C5-C6)

Hence,
the title
supinator
is not really
appropriate
for this muscle
(see p. 153).

Its contraction can be seen along the radius
when the elbow is flexed
and the forearm is pronated or supinated.

Biceps brachii has two origins.

The long head (left) arises from a tubercle above the glenoid cavity of the scapula, travels through the shoulder joint, between the greater and lesser tubercles, and along the bicipital groove before merging with the body.

The short head (right) starts as a tendon at the coracoid process on the lateral edge of the scapula and becomes a fleshy body which joins with the muscle fibers of the long head about halfway down the humerus.

The two heads then continue downward and form one tendon, which passes anterior to the elbow joint and inserts at the bicipital tuberosity of the radius.

At the shoulder level, the two heads of biceps brachii have different actions (see p. 129).

Actions: This muscle is the primary elbow flexor. It also supinates the radius at the elbow.

Contraction of this muscle is very visible on the anterior part of the arm, when the elbow is flexed and the forearm supinated.

Innervation:
musculocutaneous nerve (C5-C6)

Muscles for extension of elbow

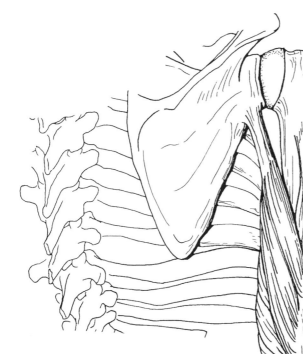

Triceps brachii, as the name suggests, consists of three heads.

The long head comes from a tubercle below the glenoid cavity of the scapula, the lateral head is from the lateral posterosuperior shaft of the humerus, and the medial head (better called the deep head) is from the posteroinferior humerus, where it is covered up by the body.

Triceps has a single broad insertion by a tendon onto the olecranon.

Action: This is the major elbow extensor.

The long head also participates in extension of the arm due to its attachment to the scapula.

Anconeus is a small muscle running from the lateral epicondyle of the humerus to the superior ulna.

Action: extends the elbow; plays a small role as an abductor during pronation of the ulna (see p. 149)

Innervation: radial nerve (C7-C8)

Innervation: radial nerve (C7-C8)

Pronation/supination of elbow and forearm

These movements involve changes in the elbow joint
and in the relationship between the radius and ulna.

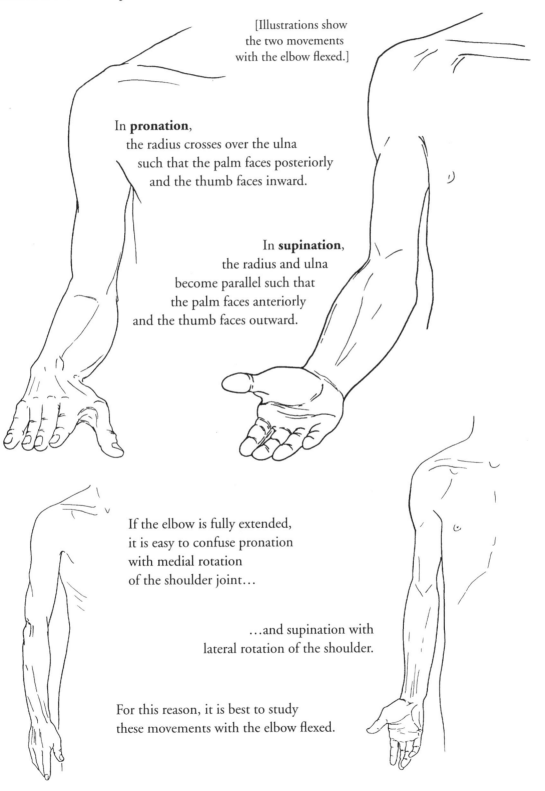

[Illustrations show
the two movements
with the elbow flexed.]

In **pronation**,
the radius crosses over the ulna
such that the palm faces posteriorly
and the thumb faces inward.

In **supination**,
the radius and ulna
become parallel such that
the palm faces anteriorly
and the thumb faces outward.

If the elbow is fully extended,
it is easy to confuse pronation
with medial rotation
of the shoulder joint…

…and supination with
lateral rotation of the shoulder.

For this reason, it is best to study
these movements with the elbow flexed.

Pronation and supination are possible because of the interplay
of articular surfaces and ligaments at the proximal and distal ends of the forearm.

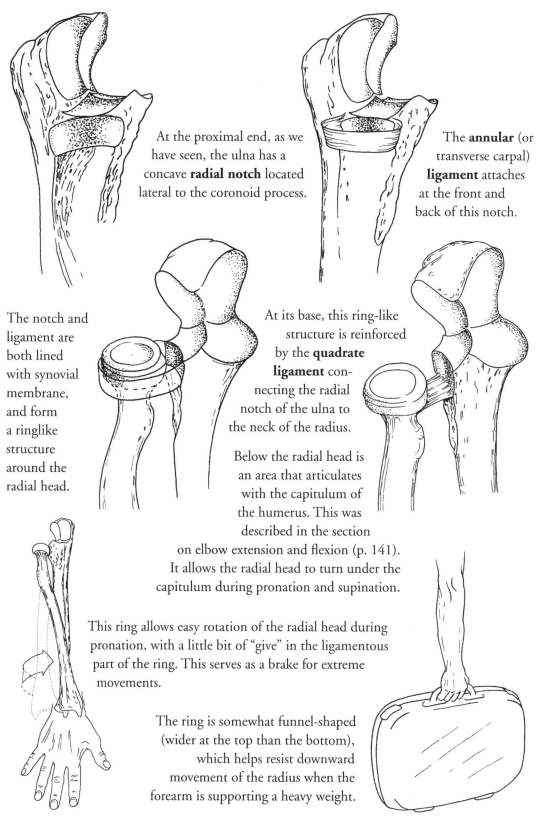

At the proximal end, as we
have seen, the ulna has a
concave **radial notch** located
lateral to the coronoid process.

The **annular** (or
transverse carpal)
ligament attaches
at the front and
back of this notch.

The notch and
ligament are
both lined
with synovial
membrane,
and form
a ringlike
structure
around the
radial head.

At its base, this ring-like
structure is reinforced
by the **quadrate
ligament** con-
necting the radial
notch of the ulna to
the neck of the radius.

Below the radial head is
an area that articulates
with the capitulum of
the humerus. This was
described in the section
on elbow extension and flexion (p. 141).
It allows the radial head to turn under the
capitulum during pronation and supination.

This ring allows easy rotation of the radial head during
pronation, with a little bit of "give" in the ligamentous
part of the ring. This serves as a brake for extreme
movements.

The ring is somewhat funnel-shaped
(wider at the top than the bottom),
which helps resist downward
movement of the radius when the
forearm is supporting a heavy weight.

At the distal end, the two bones of the forearm have several surfaces.

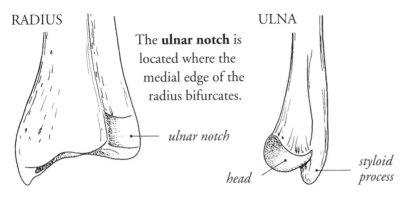

RADIUS

ULNA

The **ulnar notch** is located where the medial edge of the radius bifurcates.

— *ulnar notch*

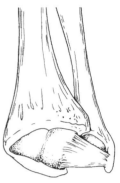

styloid process

head

The distal radioulnar joint consists of the convex ulnar head fitting into the concave ulnar notch of the radius. This facilitates rotation of the base of the radius around the head of the ulna.

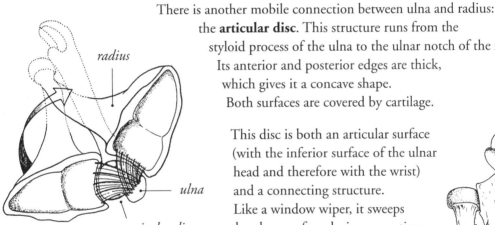

radius

ulna

articular disc

There is another mobile connection between ulna and radius: the **articular disc**. This structure runs from the styloid process of the ulna to the ulnar notch of the radius. Its anterior and posterior edges are thick, which gives it a concave shape. Both surfaces are covered by cartilage.

This disc is both an articular surface (with the inferior surface of the ulnar head and therefore with the wrist) and a connecting structure. Like a window wiper, it sweeps the ulnar surface during pronation and supination.

During pronation, the posterior portion of the disc becomes taut; during supination, the anterior portion becomes taut.

The **interosseous membrane**, which connects the shafts of the radius and ulna, has diagonal fibers oriented in both directions.

This membrane is very sturdy and consists of two layers:
• oblique middle fibers inferomedially
• oblique superior fibers superomedially.

This membrane is relaxed in pronation and taut in supination, and thus acts as a brake on supination. It also prevents longitudinal displacement of the two bones, e.g., when carrying a heavy object.

How shape of bones affects pronation/supination

Pronation involves a conical movement of the radius around the ulna. The proximal end of the radius pivots on itself, but with a slight "give" due to the suppleness of the annular ligament. Its distal end glides anteromedially around the head of the ulna.

For the ulna, there are two slightly different types of pronation:

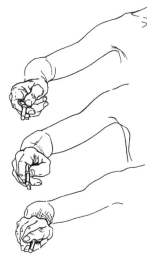

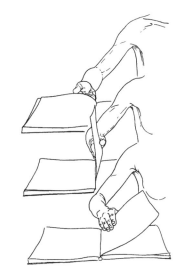

In the first (e.g., turning a key), the axis for movement of the hand passes through the middle finger, and the ulna moves slightly in conjunction with the radius. Anconeus is involved in this movement.

In the second (e.g., flipping the page of a book), the axis of the hand passes through the fifth finger, and the ulna remains fixed.

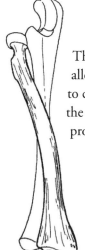

The ulna and radius, in anatomical position, are both concave anteriorly.

This curvature allows the radius to cross over the ulna during pronation.

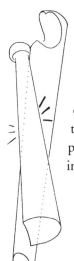

If both bones were straight, they would contact each other too soon and normal pronation would be impossible.

Fractures or other injuries can alter these curvatures and thereby interfere with pronation. This is a point of concern in certain disciplines (e.g., martial arts) involving unusual stresses on the forearm.

Pronators are attached to three bones

Humerus:

pronator teres
brachioradialis

Radius:

pronator teres
pronator quadratus
brachioradialis

Ulna:

pronator teres
pronator
  quadratus

Pronator teres arises
from two fasciae on
the medial epicondyle
of the humerus and
coronoid process of the ulna,
and inserts on the midlateral
surface of the radius.

Action:
the major pronator of the forearm,
and assists in flexion of the elbow

Innervation: median nerve (C6-C7)

Pronator quadratus
is a square-shaped muscle
running between the
anterior surfaces of the
distal ulna and radius.

Action:
pulls the radius
across the ulna
in pronation

Innervation:
anterior
interosseous
nerve
(C8-T1)

Brachioradialis
is described on page 146.
Although primarily
an elbow flexor, it can assist
in the initial stage of pronation
from a supinated position.
Conversely, it can also assist in the initial
stage of supination from a pronated position!
In other words, it tends to move the radius
to a position intermediate between
complete supination and complete pronation.

Supinators are attached to four bones

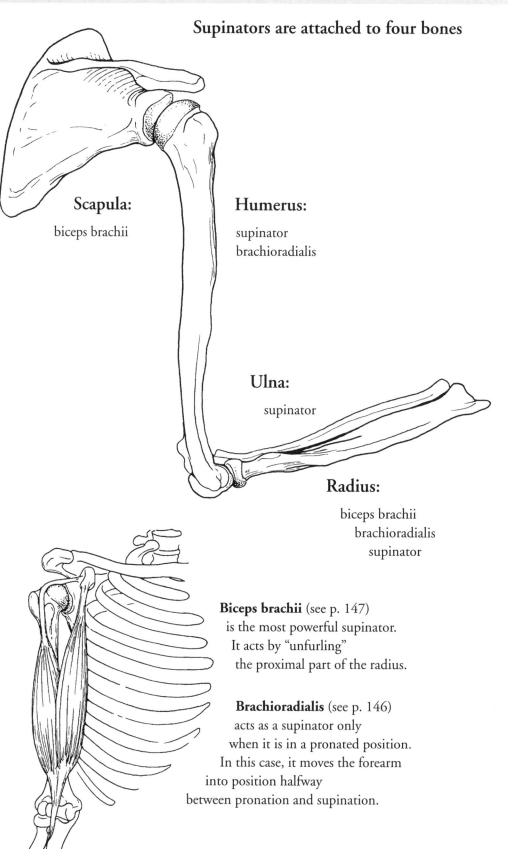

Scapula:

biceps brachii

Humerus:

supinator
brachioradialis

Ulna:

supinator

Radius:

biceps brachii
brachioradialis
supinator

Biceps brachii (see p. 147)
is the most powerful supinator.
It acts by "unfurling"
the proximal part of the radius.

Brachioradialis (see p. 146)
acts as a supinator only
when it is in a pronated position.
In this case, it moves the forearm
into position halfway
between pronation and supination.

Supinator has two layers,
a deep one (shown at left)
and a superficial one (below right).
These layers originate from
the superolateral part
of the ulna and
the lateral epicondyle
of the humerus, respectively.

This muscle wraps
around the radius,
inserting between
the neck (deep fibers),

...and the lateral side
of the bone
(superficial fibers).

Action: supinates
the forearm

Innervation: deep radial
nerve (C5–C6)

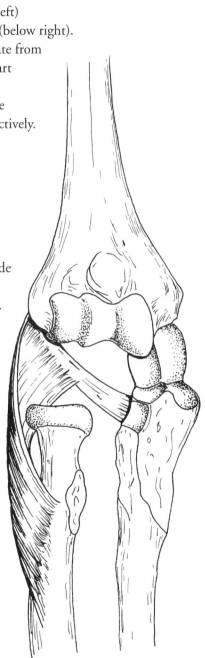

The radius can be
visualized as having
a "supinating curve"
(with biceps brachii
and supinator
inserting
near the top),

...and a "pronating curve"
(with pronator teres
inserting
near the top).

By alternately contracting,
these muscles turn the radius
like a crank.

CHAPTER FIVE
The Wrist & Hand

. . . .

The **hand**, located at the extremity of the upper body, is a very versatile tool.

This is due to the enormous *mobility* of the fingers, which are equipped with a complex system of tendons—witness the hands of a pianist, for example.

The other factor contributing to the hand's versatility is the *arrangement of the thumb* vis-à-vis the fingers. The *opposable thumbs* makes it possible to grasp objects and to accomplish many different tasks, ranging from fine precision (threading a needle) to great strength (lifting a heavy object, pulling on a partner).

The hand is linked to the forearm via the carpal bones, which form the area of the **wrist**. In this chapter, we will describe the wrist and hand together since they share many muscles.

Because the thumb plays such a major role in the bone and muscle structure of the hand, it will be dealt with separately at the end of the chapter.

Landmarks

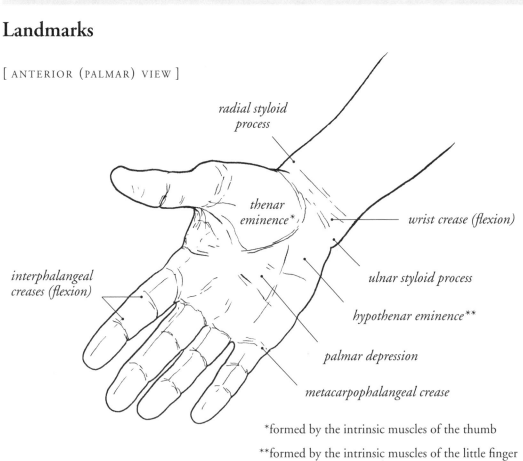

[ANTERIOR (PALMAR) VIEW]

radial styloid
process

thenar
eminence*

wrist crease (flexion)

interphalangeal
creases (flexion)

ulnar styloid process

hypothenar eminence**

palmar depression

metacarpophalangeal crease

*formed by the intrinsic muscles of the thumb

**formed by the intrinsic muscles of the little finger

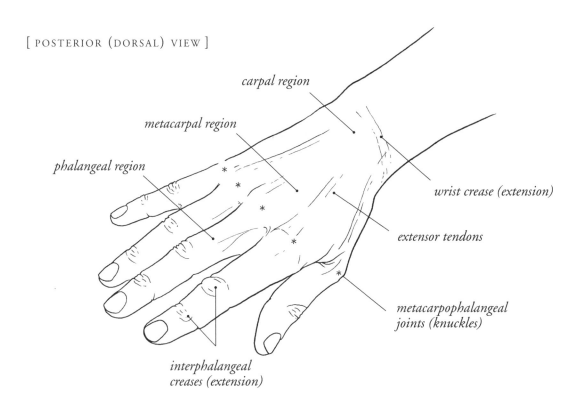

[POSTERIOR (DORSAL) VIEW]

carpal region

metacarpal region

phalangeal region

wrist crease (extension)

extensor tendons

metacarpophalangeal
joints (knuckles)

interphalangeal
creases (extension)

Bones

The skeleton of the hand, shown here with the palm facing forward,
consists of three bony areas:

At the top are the **carpal bones**, consisting of eight small bones.
The carpals are arranged in two rows of four bones each.

The first row connects
with the bones   of the forearm.

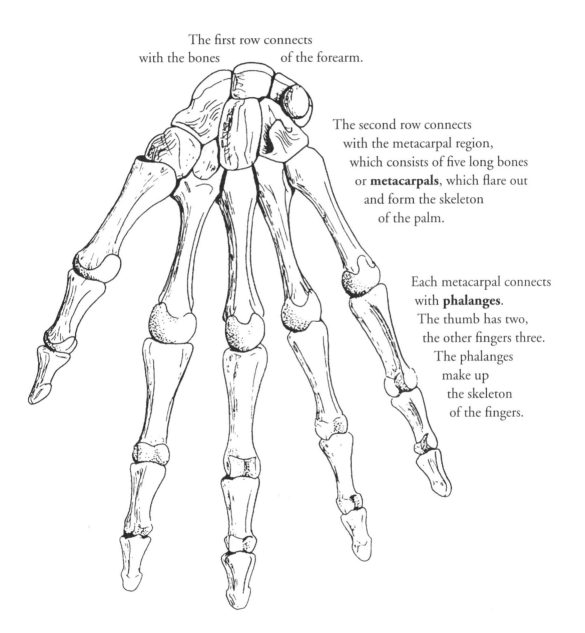

The second row connects
with the metacarpal region,
which consists of five long bones
or **metacarpals**, which flare out
and form the skeleton
of the palm.

Each metacarpal connects
with **phalanges**.
The thumb has two,
the other fingers three.
The phalanges
make up
the skeleton
of the fingers.

The metacarpals and phalanges
flare out like a fan.

Movements of the wrist

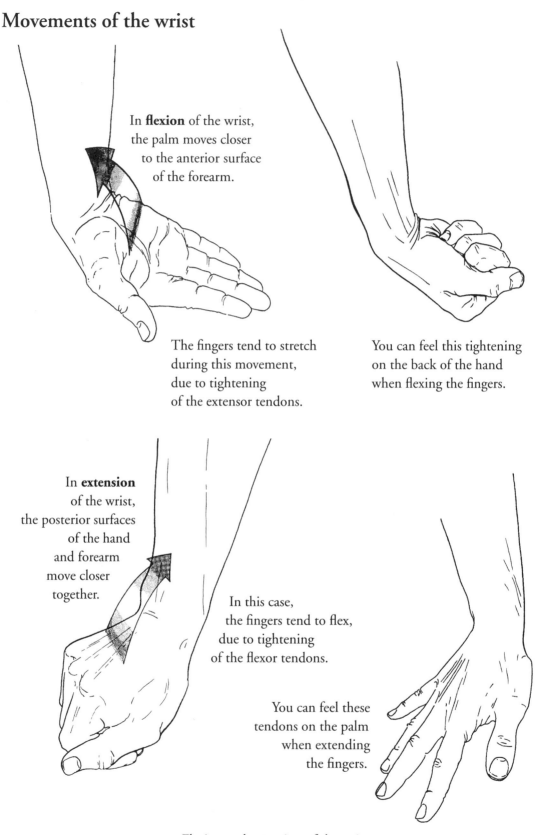

In **flexion** of the wrist, the palm moves closer to the anterior surface of the forearm.

The fingers tend to stretch during this movement, due to tightening of the extensor tendons.

You can feel this tightening on the back of the hand when flexing the fingers.

In **extension** of the wrist, the posterior surfaces of the hand and forearm move closer together.

In this case, the fingers tend to flex, due to tightening of the flexor tendons.

You can feel these tendons on the palm when extending the fingers.

Flexion and extension of the wrist have roughly the same range of motion.

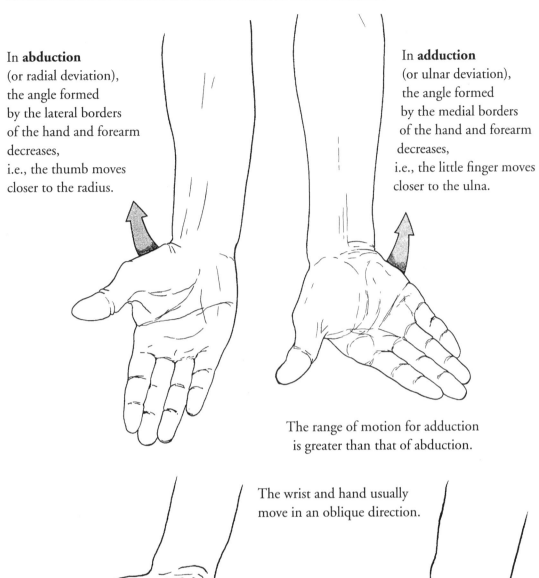

In **abduction**
(or radial deviation),
the angle formed
by the lateral borders
of the hand and forearm
decreases,
i.e., the thumb moves
closer to the radius.

In **adduction**
(or ulnar deviation),
the angle formed
by the medial borders
of the hand and forearm
decreases,
i.e., the little finger moves
closer to the ulna.

The range of motion for adduction
is greater than that of abduction.

The wrist and hand usually
move in an oblique direction.

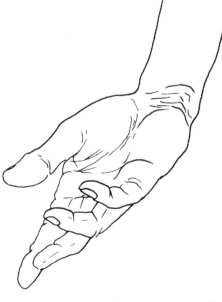

Because of the muscles involved,
adduction tends to be combined
with flexion...

...whereas abduction tends
to occur together with extension.

[Movements of the fingers are described on p. 169.]

Carpal bones

The wrist, which is only 3cm in height and 5cm in width, consists of two rows, each containing four bones.

[RIGHT HAND, PALMAR VIEW]

At the top, the "radiocarpal" row articulates with the forearm.

At the bottom, the "metacarpal" row articulates with the metacarpals.

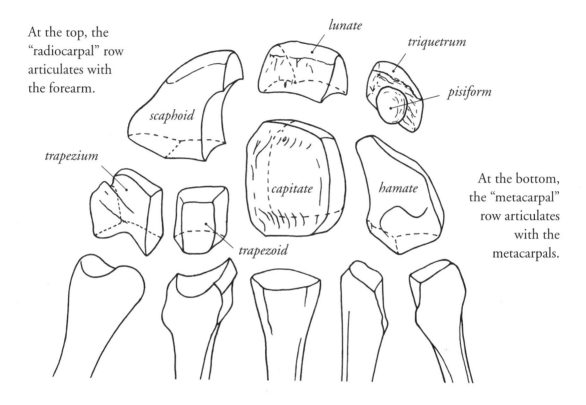

The **scaphoid** articulates superiorly with the radius and inferiorly with the trapezium and trapezoid.

The **lunate** articulates superiorly with the radius and articular disc and inferiorly with the capitate.

The **triquetrum** articulates superiorly with the articular disc. Inferiorly, it contacts the hamate and capitate.

The **pisiform** is a small round bone which sits on the anterior surface of the triquetrum. It does not articulate with the forearm nor with the hamate, but does serve for attachment of some ligaments.

The **trapezium** has a sharp anterior crest. It joins metacarpal I.

The **trapezoid** is the most symmetrical of the carpal bones, being shaped like a pyramid with the top cut off. It articulates with metacarpal II.

The **capitate** is the largest carpal and has an anterior tubercle. It articulates primarily with metacarpal III and has two facets on the inferior corners which contact metacarpals II and IV.

The **hamate** has a prominent anterior projection called the "hook." The inferior surface of the hamate has two facets oriented in different directions which articulate with metacarpals IV and V.

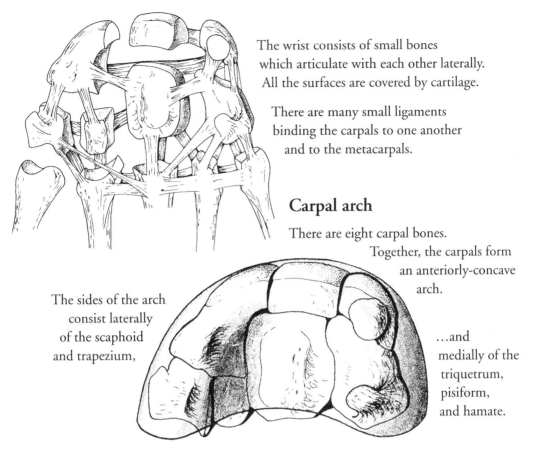

The wrist consists of small bones
which articulate with each other laterally.
All the surfaces are covered by cartilage.

There are many small ligaments
binding the carpals to one another
and to the metacarpals.

Carpal arch

There are eight carpal bones.
Together, the carpals form
an anteriorly-concave
arch.

The sides of the arch
consist laterally
of the scaphoid
and trapezium,

…and
medially of the
triquetrum,
pisiform,
and hamate.

Because of the annular (or transverse carpal)
ligament which runs anterior to the carpus,
this concave space is transformed
into a tunnel, the **carpal tunnel**.

The small intrinsic muscles of the hand and
palmaris longus attach above this ligament.

The tendons of the long muscles
of the hand pass below it.

Superiorly, the scaphoid, lunate, and
triquetrum present a large convex surface
which articulates with the smaller concave
surface presented by the distal radius and
articular disc to form an ellipsoid joint.

The posterior surface is convex
and the bones are bound together here,
like on the anterior surface, by many ligaments.

Articular surfaces of wrist joint

The wrist is an articular region consisting of many bones arranged in two rows:

- At the top, the radius and the articular disc (forming the articular radiocarpal surface) articulate with the proximal row of carpal bones (except the pisiform). This is called the **radiocarpal joint.**

- Below, this proximal row articulates with the second row of carpal bones. This is called the **midcarpal joint.**

The **articular disc** maintains the wrist structure during pronation and supination. If the wrist were to directly articulate with the two bones of the forearm, it would fold on itself during pronation.

Through the articular disc, the wrist forms a quasi-continuous surface with the radius, whether the forearm is pronated or supinated. During both movements, it "wipes" the ulnar head like a windshield wiper.

Radiocarpal joint

The surface of the radiocarpal articulation has a concave oval shape, whose posterior edge is slightly lower than the anterior edge.

Laterally, it consists of the inferior surface of the radius, and medially of the inferior surface of the articular disc, covered by cartilage.

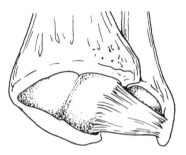

The superior surfaces of the scaphoid and lunate bones, considered together, are the **carpal condyle,** which articulates with the radius. These surfaces are covered with cartilage.

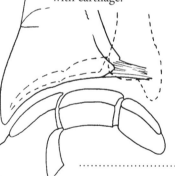

Midcarpal joint

The midcarpal joint consists of the inferior surfaces of the scaphoid, lunate, and triquetrum proximally, and the superior surfaces of trapezium, trapezoid, capitate, and hamate distally.

The space between the two rows has the shape of the letter S, where we can distinguish two parts:

- an internal part, which consists of a concave and a convex surface

- an external part, which consists of two flat surfaces superiorly and inferiorly.

Joint capsules

The radiocarpal joint is surrounded by a capsule,
which is attached to the joint circumference.
It is very loose from front to back, but taut laterally.
It is lined by a synovial membrane.
There is an articular capsule at each midcarpal joint.
The capsules are more or less joined to each other
and the synovial membrane is continuous (not shown).

Ligaments

The radiocarpal joints have many small ligaments,
which can be subdivided into three groups:

• **anterior ligaments**
run from the anterior surface of the distal radius
to the carpal bones

[FRONT VIEW]

• **lateral ligaments**
run from the styloid processes of the radius and ulna
to the carpal bones

• **posterior ligaments**
run from the posterior surface of the distal radius
and the articular disc
to the carpal bones.

[BACK VIEW]

The ligaments at the midcarpal joints
connect neighboring bones.
They are reinforced by bundles of ligaments
from the radiocarpal joint.

Movements of the wrist

The movements of the wrist involve two rows of joints.

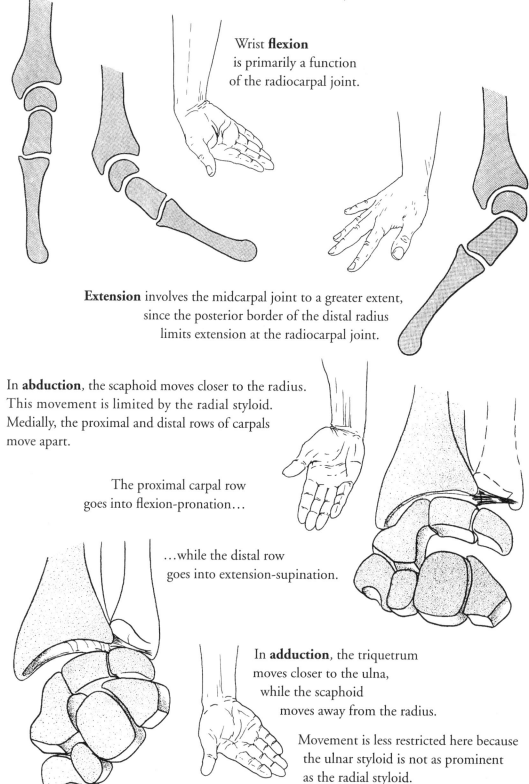

Wrist **flexion**
is primarily a function
of the radiocarpal joint.

Extension involves the midcarpal joint to a greater extent,
since the posterior border of the distal radius
limits extension at the radiocarpal joint.

In **abduction**, the scaphoid moves closer to the radius.
This movement is limited by the radial styloid.
Medially, the proximal and distal rows of carpals
move apart.

The proximal carpal row
goes into flexion-pronation…

…while the distal row
goes into extension-supination.

In **adduction**, the triquetrum
moves closer to the ulna,
while the scaphoid
moves away from the radius.

Movement is less restricted here because
the ulnar styloid is not as prominent
as the radial styloid.
The lateral carpals move apart during adduction.

Metacarpals and phalanges

There are five bony structures which consist of one metacarpal and several phalanges each (the thumb has two phalanges, the other fingers have three each).

Each of these bones consists of three parts:

base (proximal)

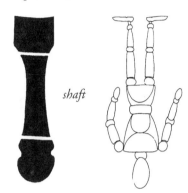

shaft

head (distal)

The bony structures studied here are those of fingers 2 through 5. The structure of the thumb is studied on page 183.

The metacarpals

The base of each **metacarpal** is roughly quadrangular, with facets for articulation with a carpal and the adjacent metacarpals.

The shaft is roughly triangular, with three surfaces and three sides.

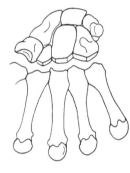

dorsal surface

two lateral sides

two lateral surfaces

palmar side

The head has one cartilaginous articulating surface, which is rounded from front to back and laterally. On each side, there is a small tubercle.

The phalanges

The **proximal phalanx** of each finger has a concavely rounded base for articulation with the metacarpal, and a pulley-shaped head.

The base of the **middle phalanx** is concave but with a median crest to match the shape of the head of the proximal phalanx. The head has the same surface as the proximal phalanx.

The base of the **distal phalanx** is identical to the base of the second phalanx. Its palmar head has a protuberance for the finger tip area.

Carpometacarpal joints (without thumb)

trapezoid   *capitate*   *hamate*

These are the joints between the distal row of carpals and the metacarpals.

The articular surfaces are straight. They allow slight sliding/gliding and flexion/extension movements.

I

II   III   IV   V

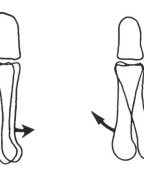

The range of these movements increases progressively from metacarpal II through V.

As a result of the anterior curvature of the carpals, the plane of carpometacarpal joints IV and V is oblique to that of joints II and III.

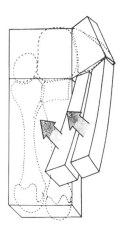

Thus, metacarpals IV and V can flex, which moves them toward the thumb. It is the carpometacarpal movements as a whole which bring about the anterior depression of the hand.

This depression is made complete by the opposition of the thumb.

Metacarpophalangeal joints

[The third finger serves as the example here.]
These are essentially hinge joints,
allowing:

Range of passive extension
is greater than that
of active extension.

flexion/
extension…

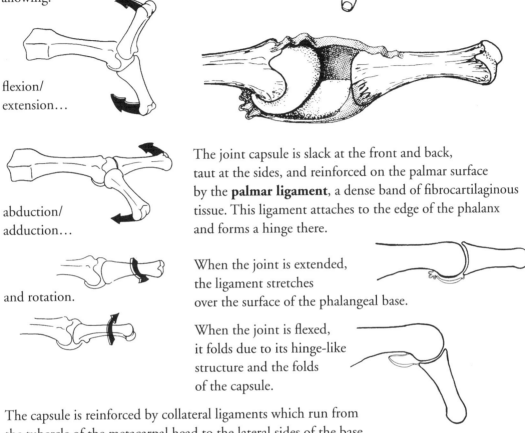

abduction/
adduction…

The joint capsule is slack at the front and back,
taut at the sides, and reinforced on the palmar surface
by the **palmar ligament**, a dense band of fibrocartilaginous
tissue. This ligament attaches to the edge of the phalanx
and forms a hinge there.

and rotation.

When the joint is extended,
the ligament stretches
over the surface of the phalangeal base.

When the joint is flexed,
it folds due to its hinge-like
structure and the folds
of the capsule.

The capsule is reinforced by collateral ligaments which run from
the tubercle of the metacarpal head to the lateral sides of the base
of the phalanx. Since they originate from the dorsal side of the metacarpal head,
which is somewhat narrower than the palmar side,
these ligaments are slack in extension and taut in flexion.

Consequently, movements of abduction/adduction and rotation
are impossible when the joint is in full flexion.

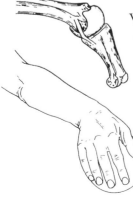

When the metacarpophalangeal joints are in
extension or slight flexion, passive abduction/
adduction and rotation allow the hand to
adapt itself to grasp a variety of shapes.

When these joints are in a more flexed
position, they become less flexible but
also more stable, which is helpful for
feats requiring strength or force.

The collateral ligaments expand like
a fan toward the palmar ligament.

Interphalangeal joints

The articular surfaces of these joints can be compared
to a convex double-track, which articulates
with a concave double-track. They allow for
anterior and posterior movements (in the sagittal plane).

The capsule and ligaments are arranged similarly
to the metacarpophalangeal joints.

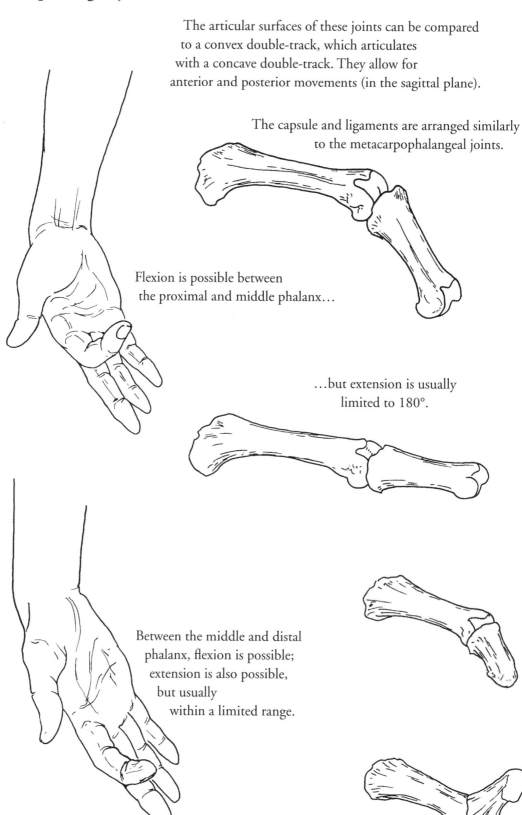

Flexion is possible between
the proximal and middle phalanx…

…but extension is usually
limited to 180°.

Between the middle and distal
phalanx, flexion is possible;
extension is also possible,
but usually
within a limited range.

Muscles of the wrist and hand with their many bony attachments

Muscles that directly move the wrist are shown in regular type.
Muscles that move the fingers and indirectly the wrist are shown in *italic*.

Humerus:

palmaris longus
flexor carpi radialis
flexor carpi ulnaris
flexor digitorum superficialis
extensor carpi radialis longus
extensor carpi radialis brevis
extensor digitorum
extensor digiti minimi
extensor carpi ulnaris

Radius:

flexor digitorum superficialis
flexor policis longus
abductor policis longus

Ulna:

flexor digitorum profundus
flexor digitorum superficialis
flexor policis longus
flexor carpi ulnaris
abductor pollicis longus
extensor pollicis longus
extensor pollicis brevis
extensor indicis
extensor carpi ulnaris

Carpal and metacarpal joints:

palmaris longus
flexor carpi radialis
flexor carpi ulnaris
extensor carpi radialis longus
extensor carpi radialis brevis
extensor carpi ulnaris
abductor pollicis longus

Phalanges:

flexor digitorum profundus
flexor digitorum superficialis
flexor policis longus
extensor pollicis longus
extensor pollicis brevis
extensor digitorum
extensor indicis
extensor digiti minimi

In addition, there are muscles that only attach to the bones of the hand, the *intrinsic muscles of the hand.* The muscles that move the thumb and form the mound on the palm at the base of the thumb are called the *thenar eminence.* The muscles that move the little finger and form a fleshy eminence on the palm along the ulnar margin are called *hypothenar eminence.* There are also intrinsic muscles located between the metacarpals, called the *interosseous* and *lumbrical* muscles.

Flexors of the wrist

There are three flexor muscles located
at the anterior surface of the forearm.
They originate at the medial epicondyle
of the humerus and insert
into the wrist.

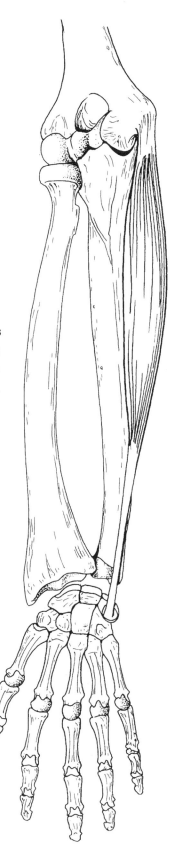

Flexor carpi ulnaris
runs from the common flexor origin
at the medial epicondyle of the humerus
and descends along the medial ulna
and styloid process,
inserting on the pisiform,
and also the hamate.

Actions: flexes and adducts the wrist;
plays a very minor role
in assisting flexion of the elbow

Innervation: ulnar nerve (C7-C8)

Palmaris longus

arises from the common flexor origin
at the medial epicondyle of the humerus
and inserts on the flexor retinaculum
and palmar aponeurosis.

Actions: flexes the wrist
and assists weakly in elbow flexion

Innervation: median nerve (C7-C8)

— palmaris longus

flexor carpi radialis

Flexor carpi radialis

arises from the common flexor origin
at the medial epicondyle of the humerus
and then runs down the forearm.
Its tendon runs through
the carpal tunnel and inserts
on the bases of metacarpals II and III.

Actions: flexes and abducts the wrist,
acting on both the radiocarpal
and midcarpal joints;
assists weakly
in elbow flexion and pronation

Innervation: median nerve (C6-C7)

Extensors of the wrist

Radialis muscles

These two muscles pass lateral to the radius, through a fibrous sheath at the level of the wrist, and end on the posterior side of the hand.

Extensor carpi radialis longus
originates from the lateral epicondyle (common extensor origin) and supracondylar ridge of the humerus.
Its tendon passes under the extensor retinaculum and inserts on the posterior base of metacarpal II.

Extensor carpi radialis brevis
arises from the common extensor origin and inserts on the posterior base of metacarpal III.

Actions: extends the wrist; participates in elbow flexion

Innervation:
radial nerve (C7-C8)

Actions: extends the wrist, abducts the hand (radial deviation); participates in elbow flexion

Innervation:
radial nerve (C6-C7)

Extensor carpi ulnaris
originates from the common extensor origin
and the posterior border of the ulna,
passes under the extensor retinaculum,
and inserts on the posterior base of metacarpal V.

Actions: extends
and adducts the wrist;
participates weakly
in elbow extension

Innervation:
radial nerve (C7-C8)

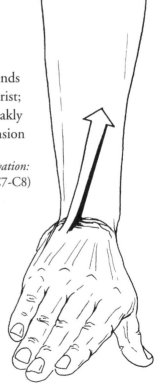

Extrinsic flexors of the fingers

These are two muscles whose mass is arranged on top of each other
on the anterior surface of the forearm and whose tendons end on the phalanges.

Flexor digitorum profundus
has a broad origin on the anterior and medial ulna,
and the medial half of the interosseous membrane
which connects the ulna and radius.
It splits into four tendons which pass through the carpal tunnel
and insert on the distal phalanges of fingers II through V.

At the level of the metacarpals,
these tendons attach to the lumbrical muscles.

Actions: flexes the third phalanx
toward the second
and participates
in flexion of the other two phalanges

Innervation: median
and ulnar nerves (C7-T1)

Once at the level
of the middle phalanx,
a tendon passes through a notch
formed when the tendon
of the flexor digitorum superficialis
splits in two.

Flexor digitorum superficialis
has two heads:
one from the common flexor origin
and the coronoid process of the ulna;
the other from the anterior surface
of the radius.

The muscle splits into four tendons
which pass
through the carpal tunnel
(superficial to the tendons
of flexor digitorum profundus,
of course), split into "Y" shapes
(to accommodate passage
of other flexor tendons),
and inserts bilaterally
on the middle phalanges
of fingers II through V.

Actions: flexes the
second phalanx toward
the first, and, due to the
fibrous sheath, flexes
the first phalanx on the
metacarpals. It assists
in flexion of the wrist,
and plays a weak role
in elbow flexion.

Innervation: median
nerve (C7-T1)

Side view of a finger, showing
the tendons of the flexor muscles

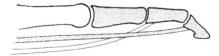

Action of flexor digitorum profundus

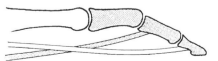

Action of flexor digitorum superficialis

Extrinsic extensors of the fingers

These three muscles are located on the posterior side of the forearm.
Their tendons insert on the posterior side of the hand.

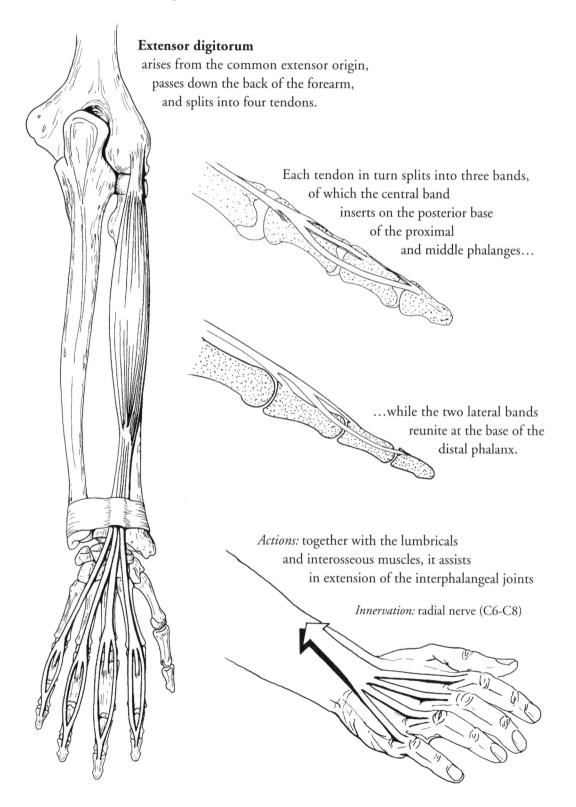

Extensor digitorum
arises from the common extensor origin,
passes down the back of the forearm,
and splits into four tendons.

Each tendon in turn splits into three bands,
of which the central band
inserts on the posterior base
of the proximal
and middle phalanges…

…while the two lateral bands
reunite at the base of the
distal phalanx.

Actions: together with the lumbricals
and interosseous muscles, it assists
in extension of the interphalangeal joints

Innervation: radial nerve (C6-C8)

Extensor indicis

arises from the posterior ulna and interosseous membrane,
below the origin of extensor pollicis longus.
Its tendon joins that of extensor digitorum
leading to the index finger.

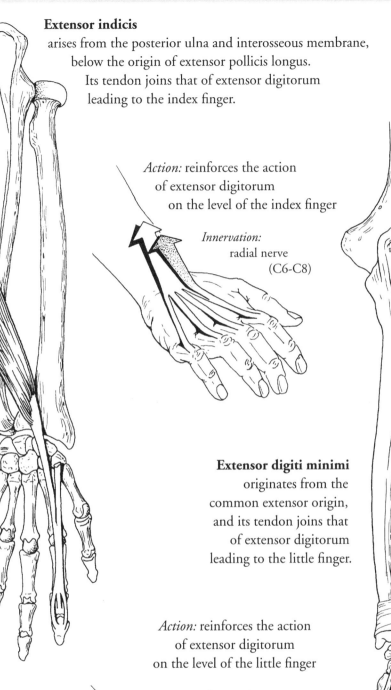

Action: reinforces the action
of extensor digitorum
on the level of the index finger

Innervation:
radial nerve
(C6-C8)

Extensor digiti minimi

originates from the
common extensor origin,
and its tendon joins that
of extensor digitorum
leading to the little finger.

Action: reinforces the action
of extensor digitorum
on the level of the little finger

Innervation:
radial nerve (C6-C8)

Intrinsic muscles that move the fingers

The intrinsic muscles of the hand are those that attach solely to the bones of the hand.

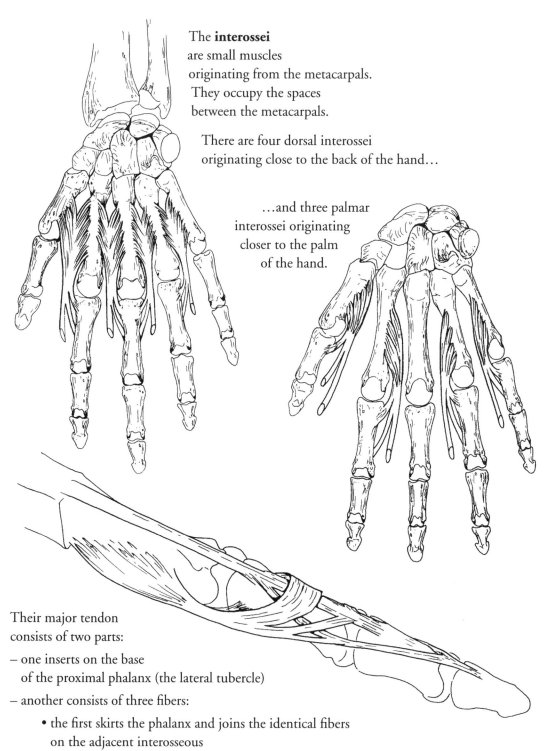

The **interossei**
are small muscles
originating from the metacarpals.
They occupy the spaces
between the metacarpals.

There are four dorsal interossei
originating close to the back of the hand…

…and three palmar
interossei originating
closer to the palm
of the hand.

Their major tendon
consists of two parts:

– one inserts on the base
   of the proximal phalanx (the lateral tubercle)
– another consists of three fibers:

   • the first skirts the phalanx and joins the identical fibers
      on the adjacent interosseous
   • the second and third insert on the edges of the extensor digitorum tendon
      at the level of the proximal and middle phalanges.

Actions: the dorsal interossei abduct
and the palmar interossei adduct
the fingers

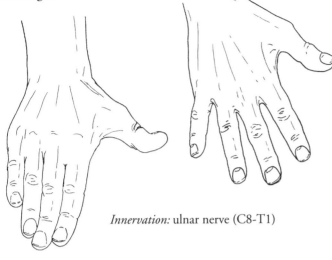

Innervation: ulnar nerve (C8-T1)

Acting together bilaterally, the first part of the tendon and the first fiber of
the second part flex the proximal phalanx. The two fibers that insert on the
extensor tendon pull on the proximal phalanx and extend the middle phalanx
toward the proximal, and the distal phalanx toward the middle.

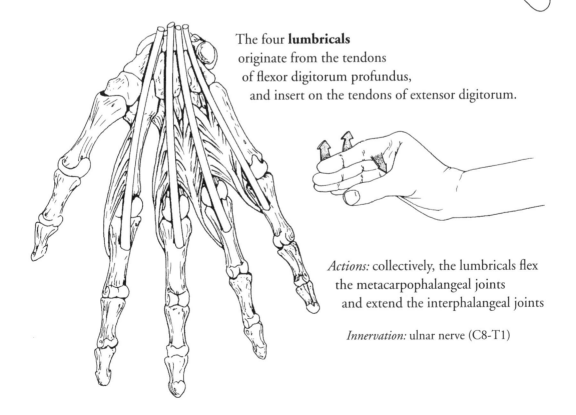

The four **lumbricals**
 originate from the tendons
  of flexor digitorum profundus,
   and insert on the tendons of extensor digitorum.

Actions: collectively, the lumbricals flex
the metacarpophalangeal joints
and extend the interphalangeal joints

Innervation: ulnar nerve (C8-T1)

Intrinsic muscles of 5th finger

The bodies of the **hypothenar muscles** provide the bulk of the hypothenar eminence on the medial side of the palm.

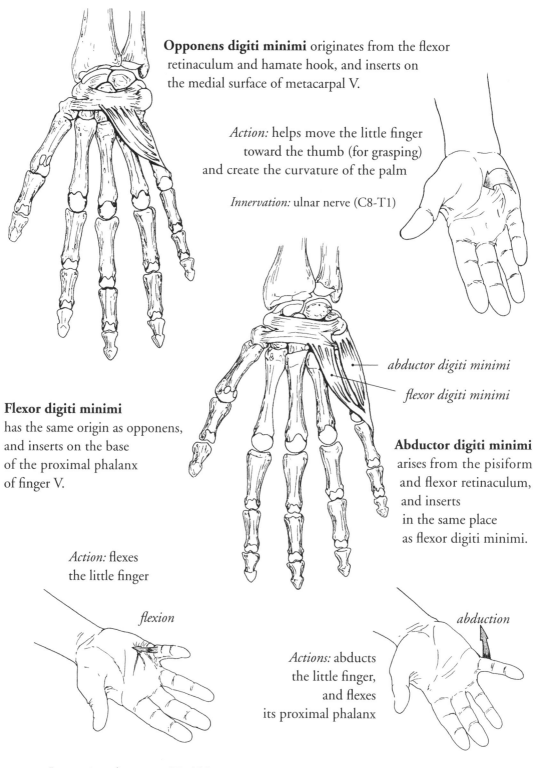

Opponens digiti minimi originates from the flexor retinaculum and hamate hook, and inserts on the medial surface of metacarpal V.

Action: helps move the little finger toward the thumb (for grasping) and create the curvature of the palm

Innervation: ulnar nerve (C8-T1)

abductor digiti minimi

flexor digiti minimi

Flexor digiti minimi has the same origin as opponens, and inserts on the base of the proximal phalanx of finger V.

Abductor digiti minimi arises from the pisiform and flexor retinaculum, and inserts in the same place as flexor digiti minimi.

Action: flexes the little finger

flexion

Actions: abducts the little finger, and flexes its proximal phalanx

abduction

Innervation: ulnar nerve (C8-T1)

Innervation: ulnar nerve (C8-T1)

Carpometacarpal articulation of thumb

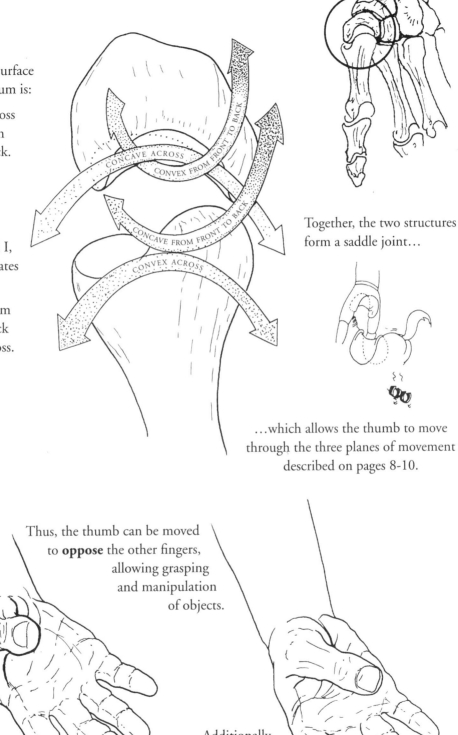

The inferior surface of the trapezium is:

– concave across
– convex from front to back.

CONCAVE ACROSS
CONVEX FROM FRONT TO BACK

The superior surface of metacarpal I, which articulates with it, is:

– concave from front to back
– convex across.

CONCAVE FROM FRONT TO BACK
CONVEX ACROSS

Together, the two structures form a saddle joint…

…which allows the thumb to move through the three planes of movement described on pages 8-10.

Thus, the thumb can be moved to **oppose** the other fingers, allowing grasping and manipulation of objects.

Additionally, the thumb has the same mobility in its metacarpophalangeal and interphalangeal joints as the other fingers.

Thumb

The thumb has a specific orientation
vis-à-vis the rest of the hand:

• the scaphoid bone
is positioned at a 40° angle
anteriorly to the carpal plane

• the first metacarpal
is positioned at a 20° angle
to the second metacarpal
and placed at a 40° angle
anteriorly.

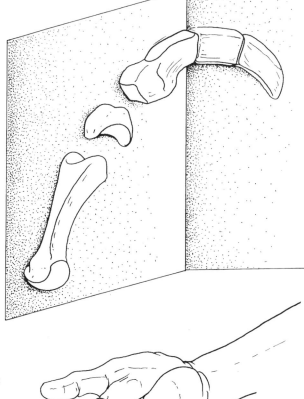

Thus, in a hand at rest, the thumb faces
the other fingers at a right angle.

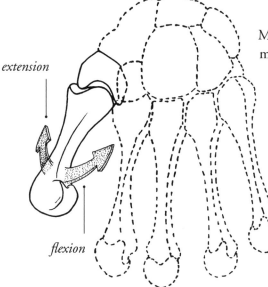

extension

flexion

Movements of the thumb (i.e., metacarpal I)
must be defined differently
from those of the other fingers.

In **extension,**
the metacarpal moves posterolaterally,
while in **flexion**
it moves anteromedially, closer to the palm.

In **abduction,**
it moves anterolaterally,
while in **adduction**
it moves posteromedially.

The capsule of carpometacarpal joint I
is slack, allowing some axial rotation
in addition to the movements
described above,
and further enhancing
the thumb's mobility.

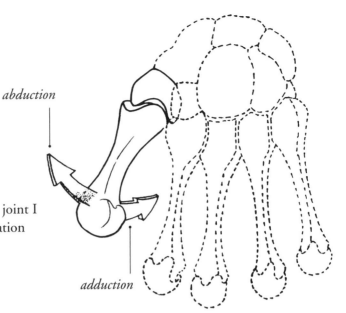

abduction

adduction

Thumb joints

Metacarpophalangeal joint I differs from II through V
in a few respects:

- It is more massive.

- The capsule is not as taut and allows some axial rotation.

- Two small sesamoid bones are embedded in the palmar fascia,
  and serve for tendon attachment.

The **interphalangeal joint** is similar to those of fingers II through V,
except for being more massive.

Extrinsic muscles of the thumb

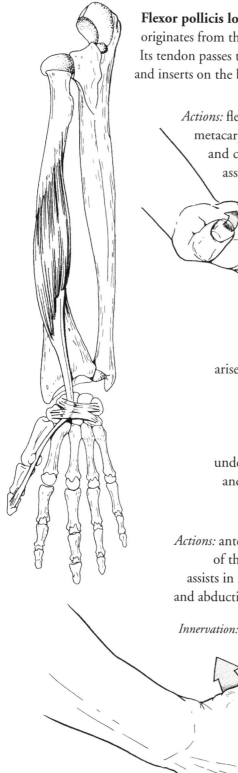

Flexor pollicis longus
originates from the anterior radius.
Its tendon passes through the carpal tunnel
and inserts on the base of the distal phalanx of the thumb.

Actions: flexion of interphalangeal joint I,
metacarpophalangeal joint I,
and carpometacarpal joint I;
assists in flexion of the wrist and abduction
(radial deviation)

Innervation: anterior interosseous
nerve (C7-C8)

Abductor pollicis longus
arises from the posterior surfaces
of the ulna, radius,
and interosseous ligament,
inferior to supinator.
The tendon passes
under the extensor retinaculum
and inserts on the lateral base
of metacarpal I.

Actions: anteromedial movement
of the thumb; also
assists in flexion of the wrist
and abduction (radial deviation)

Innervation: radial nerve (C7-C8)

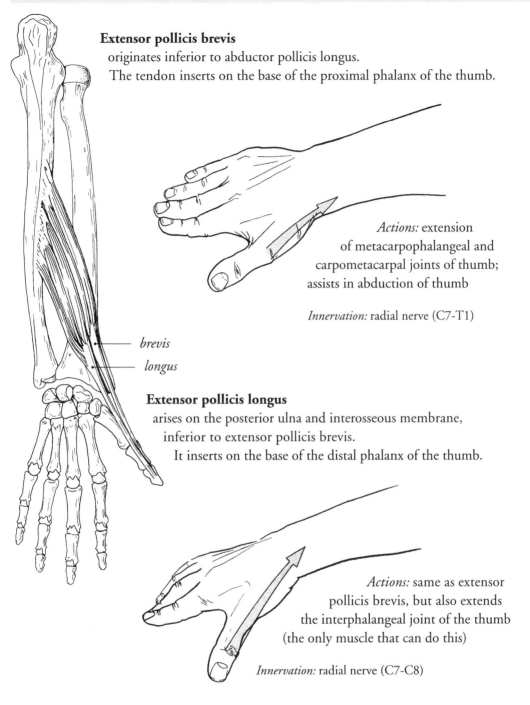

Extensor pollicis brevis

originates inferior to abductor pollicis longus.
The tendon inserts on the base of the proximal phalanx of the thumb.

Actions: extension
of metacarpophalangeal and
carpometacarpal joints of thumb;
assists in abduction of thumb

Innervation: radial nerve (C7-T1)

brevis

longus

Extensor pollicis longus

arises on the posterior ulna and interosseous membrane,
inferior to extensor pollicis brevis.
It inserts on the base of the distal phalanx of the thumb.

Actions: same as extensor
pollicis brevis, but also extends
the interphalangeal joint of the thumb
(the only muscle that can do this)

Innervation: radial nerve (C7-C8)

When the thumb is fully extended, a depression known
as the "anatomical snuffbox" can be seen
at the posterior base of the thumb.

It is bordered laterally by the
tendons of abductor pollicis longus
and extensor pollicis brevis,
and medially by the tendon of extensor pollicis longus.

Intrinsic muscles of the thumb

Adductor pollicis has two fibers:

- Adductor pollicis obliquus arises from
  the trapezoid and capitate bones.

- Adductor pollicis transversus arises from
  the 2nd and 3rd metacarpals
  and the corresponding metacarpophalangeal joint.

The two fibers insert on the medial base
of the proximal phalanx of the thumb,
and the medial sesamoid bone
located at metacarpophalangeal joint I.

Actions: moves
the 2nd metacarpal
toward the 1st;
also flexes
the metacarpophalangeal joint

Innervation: ulnar nerve (C8-T1)

Flexor pollicis brevis consists
of two layers:

- a deep layer, which arises from the trapezoid
  and capitate

- a superficial layer, which arises from the trapezium
  and the flexor retinaculum.

The two layers merge into one tendon,
which inserts on the lateral base
of the proximal phalanx of the thumb,
and the lateral sesamoid bone located
at metacarpophalangeal joint I.

Actions: moves the metacarpals anteromedially
and in medial rotation,
and flexes the proximal phalanx of the thumb

Innervation: median and ulnar nerves (C8-T1)

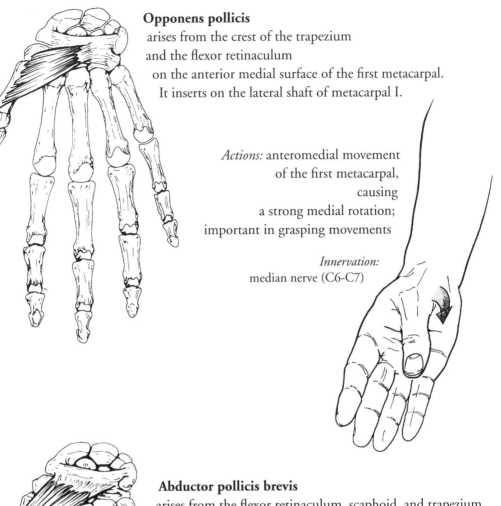

Opponens pollicis
arises from the crest of the trapezium
and the flexor retinaculum
on the anterior medial surface of the first metacarpal.
It inserts on the lateral shaft of metacarpal I.

Actions: anteromedial movement
of the first metacarpal,
causing
a strong medial rotation;
important in grasping movements

Innervation:
median nerve (C6-C7)

Abductor pollicis brevis
arises from the flexor retinaculum, scaphoid, and trapezium,
and inserts on the lateral base
of the proximal phalanx of the thumb
next to flexor pollicis brevis.

Actions:
pulls the metacarpal anteriorly
and flexes
the metacarpophalangeal joint

Innervation:
median nerve (C8-T1)

CHAPTER SIX

The Hip & Knee

. . . .

The **hip** is the proximal joint of the lower limbs and connects the femur to the pelvis. It is surrounded by thick muscles, and therefore difficult to palpate or localize.

The stability and powerful musculature of this joint are essential for standing and walking. Many physical disciplines require good range of motion (ROM) at the hip. Restrictions in ROM here are common, and typically affect nearby structures such as the lower back, knee, and foot joints. To work with the hip, we therefore need to understand this joint in order to more easily isolate its movements.

The **knee**, an intermediate joint on the lower limb, is primarily capable of flexion and extension. Its mobility allows it to vary the distance between the foot and the trunk to a large degree. Its stability is not due to bone structure, which is rather weak, but rather to the arrangement of ligaments and muscles. The knee receives considerable stress, both from above (body weight) and below (impact of the foot on the ground, footwear).

This chapter looks at the hip and knee together, because these two joints have many muscles in common.

Landmarks

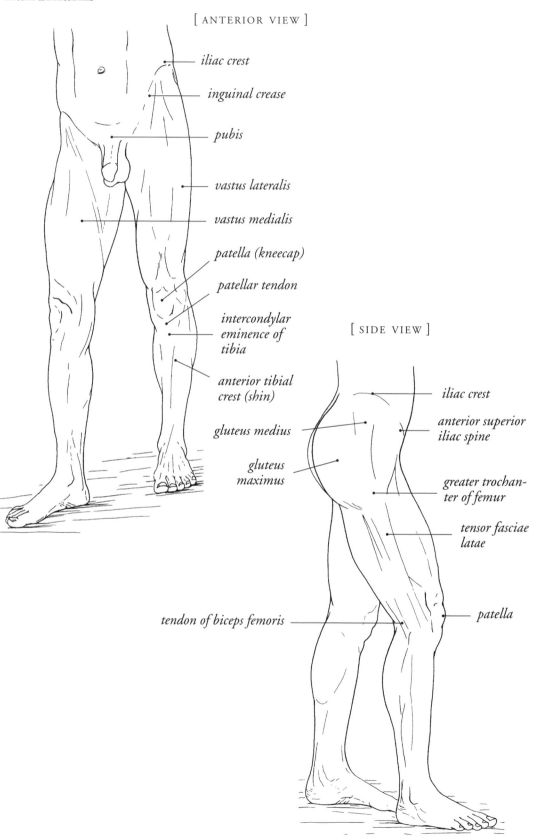

[ANTERIOR VIEW]

iliac crest

inguinal crease

pubis

vastus lateralis

vastus medialis

patella (kneecap)

patellar tendon

intercondylar
eminence of
tibia

anterior tibial
crest (shin)

gluteus medius

gluteus
maximus

tendon of biceps femoris

[SIDE VIEW]

iliac crest

anterior superior
iliac spine

greater trochan-
ter of femur

tensor fasciae
latae

patella

[POSTERIOR VIEW]

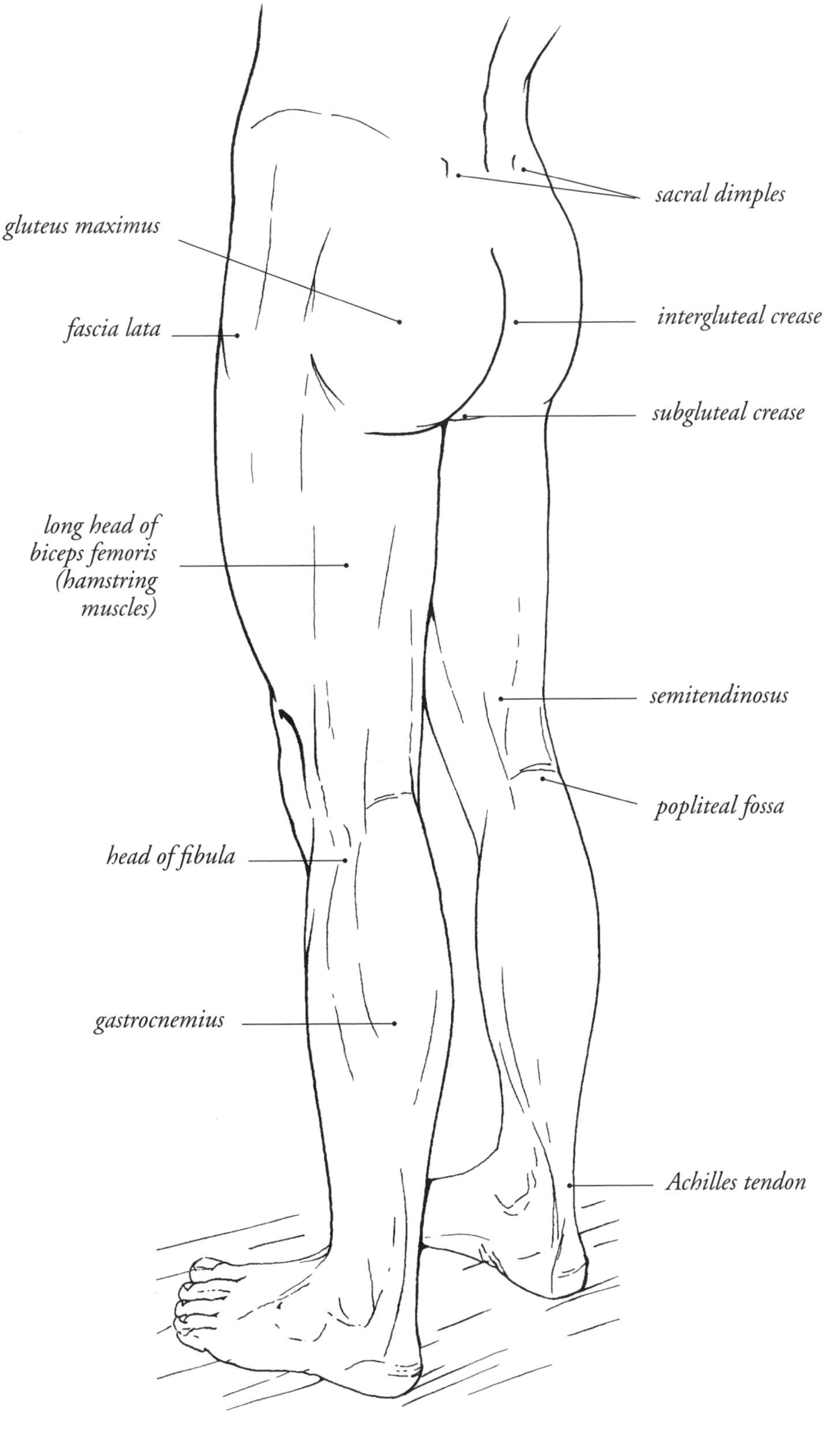

sacral dimples

gluteus maximus

intergluteal crease

fascia lata

subgluteal crease

long head of
biceps femoris
(hamstring
muscles)

semitendinosus

popliteal fossa

head of fibula

gastrocnemius

Achilles tendon

Movements of hip

The shape of its articulations (see p. 201-202)
allows the hip joint to be moved in many different directions.
For ease of study, the movements of the hip are described
with respect to the planes they intersect (see p. 8-10).

We assume first that the pelvis is fixed
and the femur is moving.

Flexion is the movement in which the
angle between the anterior surfaces of
the thigh and the trunk decreases.
ROM for hip flexion is greater
when the knee is also flexed.

ROM for passive flexion is greater
than for active flexion
since the flexing muscles are relaxed
and can be compressed.

When the knee is extended,
hip flexion is restricted
by the limits of elasticity
of the hamstring muscles (see p. 242).

Hip flexion is often associated
with extension of the pelvis.

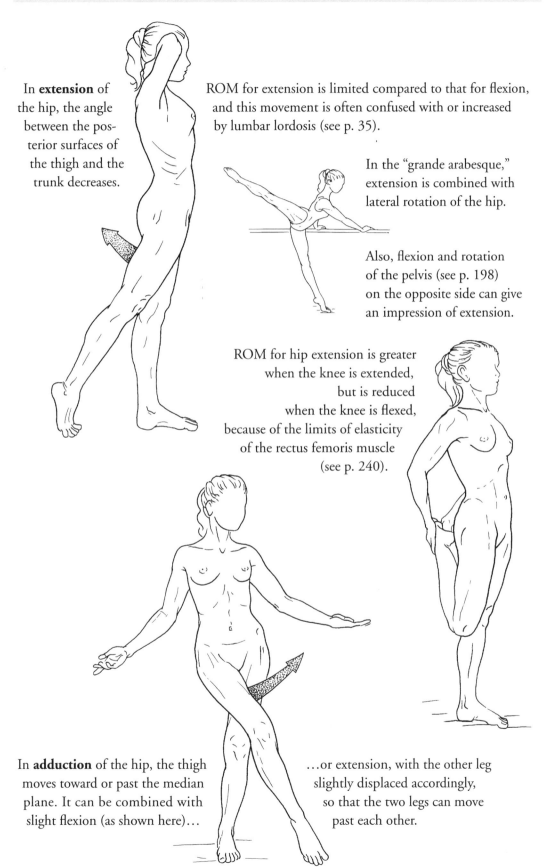

In **extension** of the hip, the angle between the posterior surfaces of the thigh and the trunk decreases.

ROM for extension is limited compared to that for flexion, and this movement is often confused with or increased by lumbar lordosis (see p. 35).

In the "grande arabesque," extension is combined with lateral rotation of the hip.

Also, flexion and rotation of the pelvis (see p. 198) on the opposite side can give an impression of extension.

ROM for hip extension is greater when the knee is extended, but is reduced when the knee is flexed, because of the limits of elasticity of the rectus femoris muscle (see p. 240).

In **adduction** of the hip, the thigh moves toward or past the median plane. It can be combined with slight flexion (as shown here)…

…or extension, with the other leg slightly displaced accordingly, so that the two legs can move past each other.

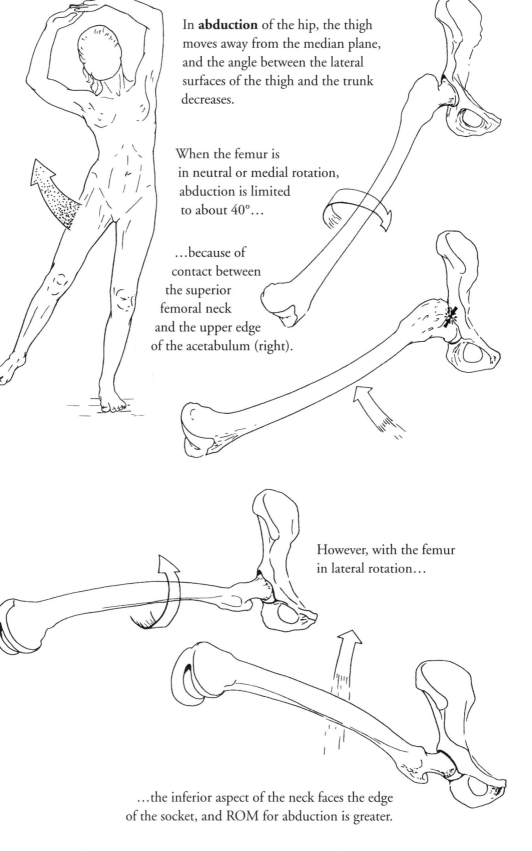

In **abduction** of the hip, the thigh moves away from the median plane, and the angle between the lateral surfaces of the thigh and the trunk decreases.

When the femur is in neutral or medial rotation, abduction is limited to about 40°…

…because of contact between the superior femoral neck and the upper edge of the acetabulum (right).

However, with the femur in lateral rotation…

…the inferior aspect of the neck faces the edge of the socket, and ROM for abduction is greater.

In **medial rotation** of the hip, the femur rotates on its own long axis, and the toes of the foot (or an imaginary spot on the front of the thigh) move closer to the median plane.

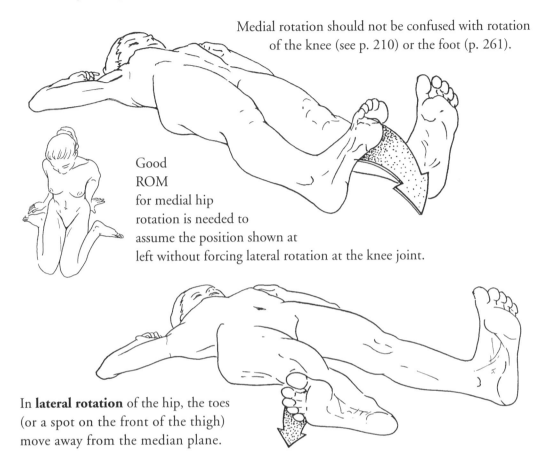

Medial rotation should not be confused with rotation of the knee (see p. 210) or the foot (p. 261).

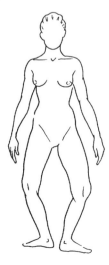

Good ROM for medial hip rotation is needed to assume the position shown at left without forcing lateral rotation at the knee joint.

In **lateral rotation** of the hip, the toes (or a spot on the front of the thigh) move away from the median plane.

Good ROM for lateral rotation is needed for the "en dehors" position of ballet...

...or for assuming the "lotus position" without stressing the knee and ankle joints.

When the hip is flexed, ROM for lateral rotation is greater because the iliofemoral ligament is slack.

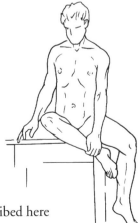

Most often, the hip movements described here combine several directions in one movement, e.g., abduction + lateral rotation, or flexion + abduction.

Let us now consider the possible movements *of the pelvis* at the hip joint, assuming that the femur is fixed. We will focus on the anterior superior iliac spine (ASIS) as a reference point.

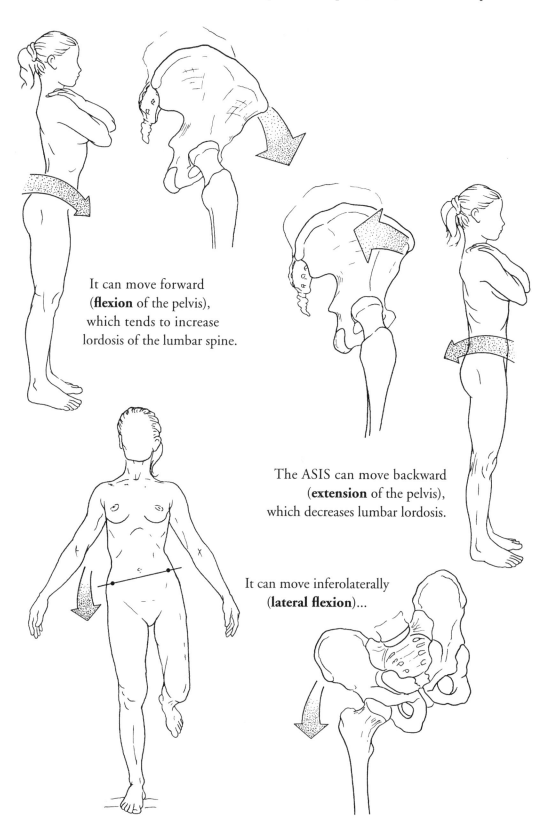

It can move forward (**flexion** of the pelvis), which tends to increase lordosis of the lumbar spine.

The ASIS can move backward (**extension** of the pelvis), which decreases lumbar lordosis.

It can move inferolaterally (**lateral flexion**)...

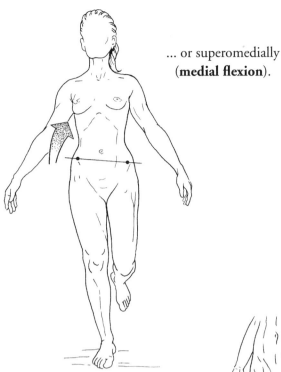

... or superomedially
(**medial flexion**).

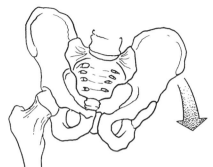

When done
in an upright position,
these two movements
are associated
with sidebending
of the lumbar spine.

Note that in these
two illustrations,
you can observe the
movement of the pelvis
on the supported hip
and not on the
unsupported hip.

Finally, the ASIS can undergo limited
medial rotation of the pelvis...

or **lateral rotation**
of the pelvis.

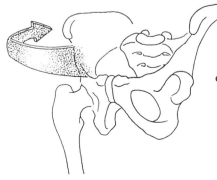

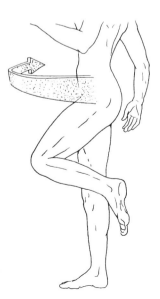

Femur

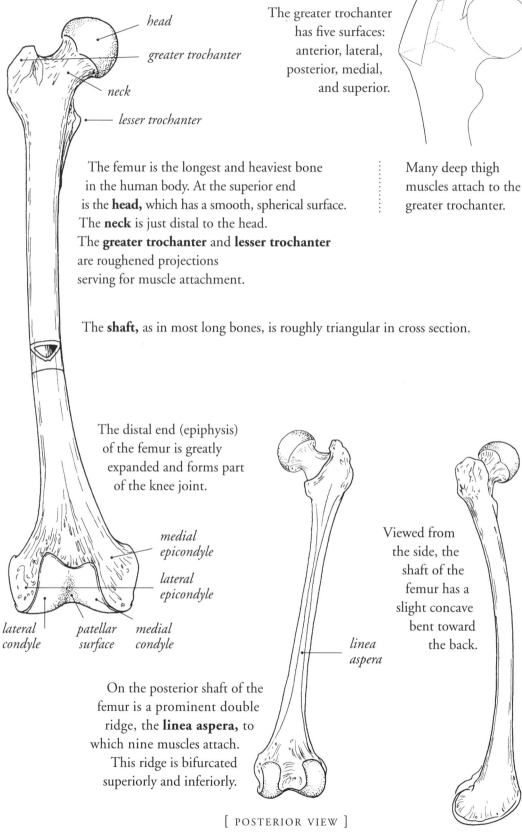

head

greater trochanter

neck

lesser trochanter

The greater trochanter has five surfaces: anterior, lateral, posterior, medial, and superior.

The femur is the longest and heaviest bone in the human body. At the superior end is the **head,** which has a smooth, spherical surface. The **neck** is just distal to the head. The **greater trochanter** and **lesser trochanter** are roughened projections serving for muscle attachment.

Many deep thigh muscles attach to the greater trochanter.

The **shaft,** as in most long bones, is roughly triangular in cross section.

The distal end (epiphysis) of the femur is greatly expanded and forms part of the knee joint.

medial epicondyle

lateral epicondyle

lateral condyle   *patellar surface*   *medial condyle*

Viewed from the side, the shaft of the femur has a slight concave bent toward the back.

linea aspera

On the posterior shaft of the femur is a prominent double ridge, the **linea aspera,** to which nine muscles attach. This ridge is bifurcated superiorly and inferiorly.

[POSTERIOR VIEW]

Hip (coxo-femoral) joint

On the lateral surface of the hip bones is the acetabulum (Latin for "small bowl"), a deep socket formed by the junction of the ilium, pubis, and ischium (see p. 44). It should not be confused with the obturator foramen.

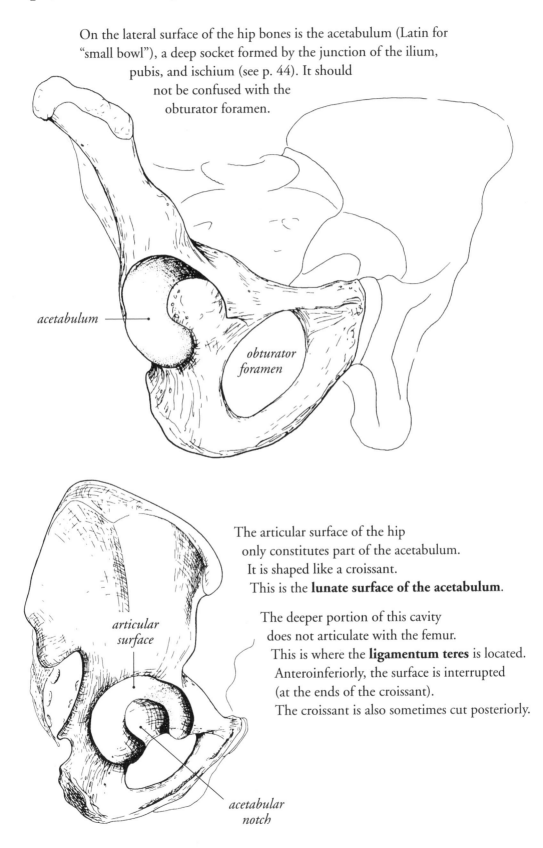

acetabulum

obturator foramen

articular surface

acetabular notch

The articular surface of the hip only constitutes part of the acetabulum. It is shaped like a croissant. This is the **lunate surface of the acetabulum**.

The deeper portion of this cavity does not articulate with the femur. This is where the **ligamentum teres** is located. Anteroinferiorly, the surface is interrupted (at the ends of the croissant). The croissant is also sometimes cut posteriorly.

Articular surfaces of hip

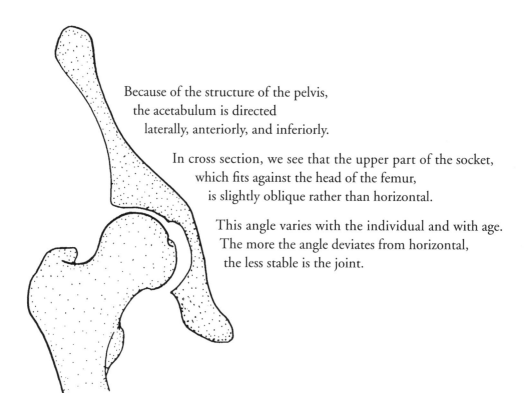

Because of the structure of the pelvis,
the acetabulum is directed
laterally, anteriorly, and inferiorly.

In cross section, we see that the upper part of the socket,
which fits against the head of the femur,
is slightly oblique rather than horizontal.

This angle varies with the individual and with age.
The more the angle deviates from horizontal,
the less stable is the joint.

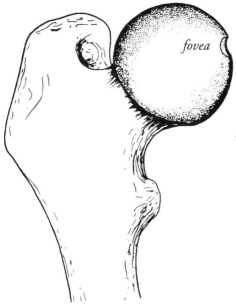

fovea

The femoral head,
the articulating surface of the femur,
represents about two-thirds of a sphere
and is covered with thick cartilage
except at the fovea
where the **ligamentum teres**,
which connects the femoral head
to the acetabulum, is attached.

The head of the femur sits on top of the **femoral neck**.

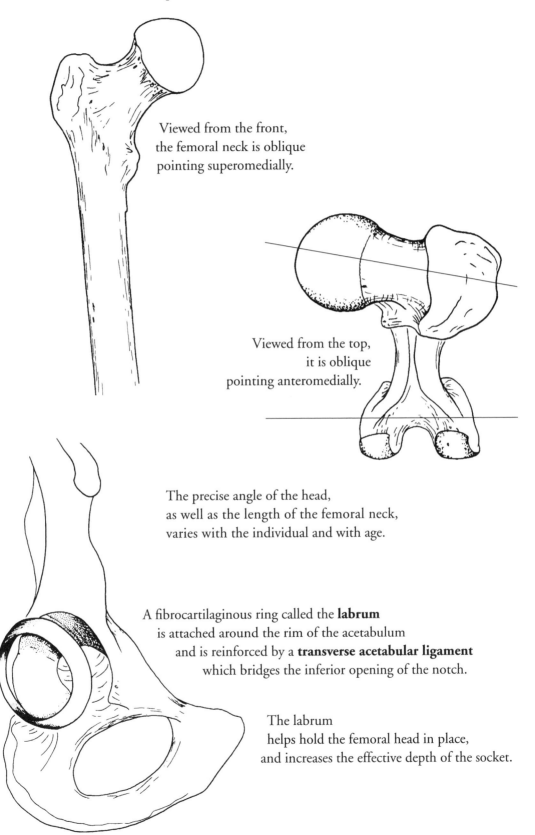

Viewed from the front,
the femoral neck is oblique
pointing superomedially.

Viewed from the top,
it is oblique
pointing anteromedially.

The precise angle of the head,
as well as the length of the femoral neck,
varies with the individual and with age.

A fibrocartilaginous ring called the **labrum**
is attached around the rim of the acetabulum
and is reinforced by a **transverse acetabular ligament**
which bridges the inferior opening of the notch.

The labrum
helps hold the femoral head in place,
and increases the effective depth of the socket.

How head of femur articulates with socket

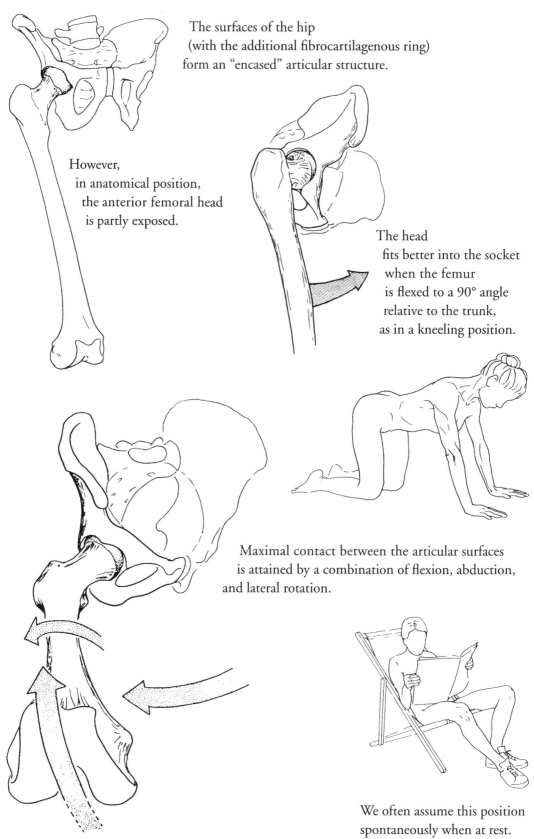

The surfaces of the hip
(with the additional fibrocartilagenous ring)
form an "encased" articular structure.

However,
in anatomical position,
the anterior femoral head
is partly exposed.

The head
fits better into the socket
when the femur
is flexed to a 90° angle
relative to the trunk,
as in a kneeling position.

Maximal contact between the articular surfaces
is attained by a combination of flexion, abduction,
and lateral rotation.

We often assume this position
spontaneously when at rest.

Variations of hip

 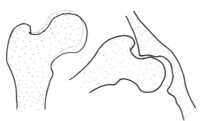 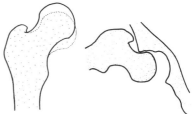

The average angle between the femoral neck and shaft is 135°.

In some individuals this angle is smaller, a condition called **coxa vara** in which the range of abduction is reduced.

When the angle is greater than 135° (**coxa valga**), the range of abduction is increased.

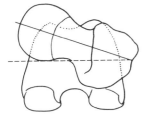 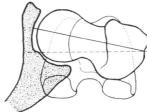 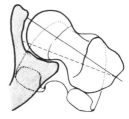

Seen from above, the neck is oriented anteriorly at an angle of 10 to 30°.

When this "anteversion" angle is small, the head fits into the socket well in anatomical position,

and maintains good articular contact even in lateral rotation.

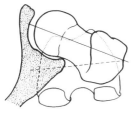 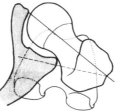 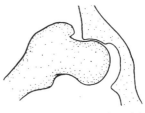

When the anteversion angle is large, the anterior part of the head is more exposed in anatomical position,

and the posterior part loses contact with the socket in lateral rotation.

Lateral rotation is more restricted in these individuals by contact between the neck and the lateral edge of the acetabulum. Curvature and length of the femoral neck also affect mobility at the hip joint.

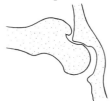 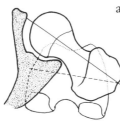 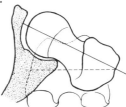

A neck that is more concave, and longer, will facilitate abduction...

and lateral rotation.

With a shorter, less concave neck, both these motions are restricted by contact with the edge of the acetabulum.

Obviously, there are intrinsic limitations to ROM at the hip joint due to the shape of the articulating bones, which may be greater or less in specific individuals. It is important to be aware of this variation when teaching dance or other physical disciplines. In fact, people with limited movement of the hip joint due to bony restrictions may injure themselves while trying to achieve certain positions, by putting too much stress on the superior joints (lumbar spine) or inferior joints (knee).

Capsule and ligaments of hip joint

The capsule of the hip joint attaches firmly
all the way around the rim of the acetabulum
and at the base of the femoral neck.

It is thick and tough, and is reinforced by ligaments.

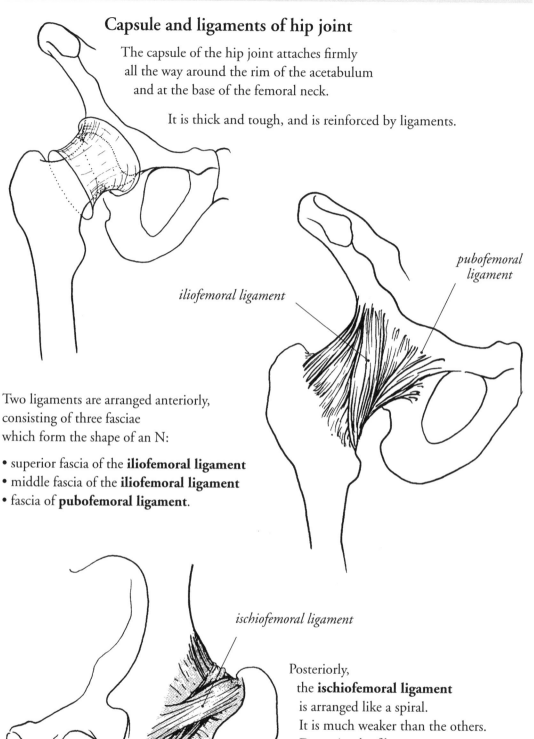

*pubofemoral
ligament*

iliofemoral ligament

Two ligaments are arranged anteriorly,
consisting of three fasciae
which form the shape of an N:

• superior fascia of the **iliofemoral ligament**
• middle fascia of the **iliofemoral ligament**
• fascia of **pubofemoral ligament**.

ischiofemoral ligament

Posteriorly,
the **ischiofemoral ligament**
is arranged like a spiral.
It is much weaker than the others.
Deep circular fibers
reinforce the middle of the capsule,
which give it an hourglass shape.

During movements of the hip, the anterior ligaments display varying degrees of tightness.

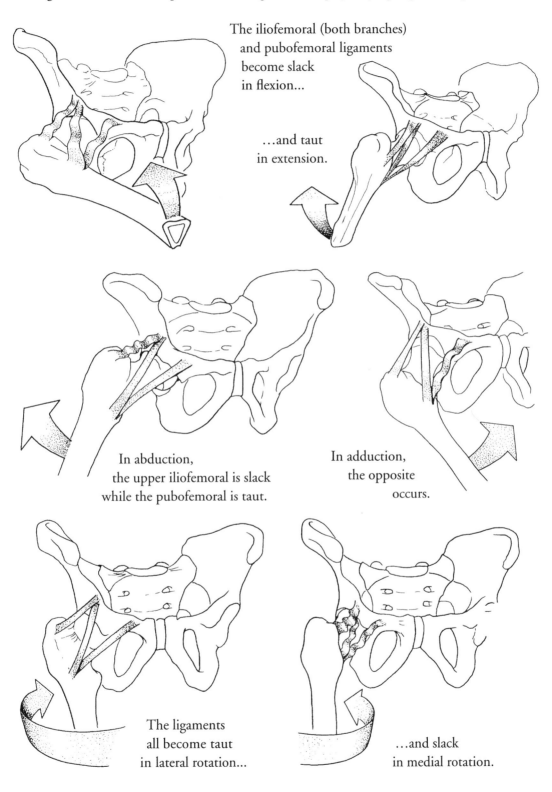

The iliofemoral (both branches) and pubofemoral ligaments become slack in flexion...

...and taut in extension.

In abduction, the upper iliofemoral is slack while the pubofemoral is taut.

In adduction, the opposite occurs.

The ligaments all become taut in lateral rotation...

...and slack in medial rotation.

In summary, flexion and medial rotation loosen the ligaments, while extension and lateral rotation make them taut.

Movements of the knee

The knee is primarily a hinge joint.

Flexion decreases the angle formed
by the posterior thigh and leg.

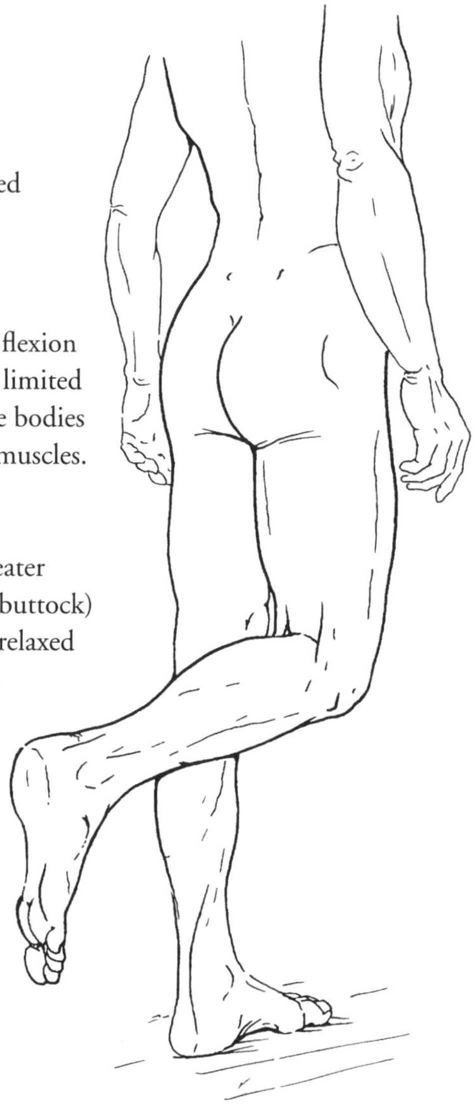

ROM for active flexion
is limited
by contact between the bodies
of the contracting muscles.

ROM for passive flexion is greater
(i.e., the heel can touch the buttock)
since the flexor muscles are relaxed
and more easily compressed.
(The extensor muscles are
passively stretched.)

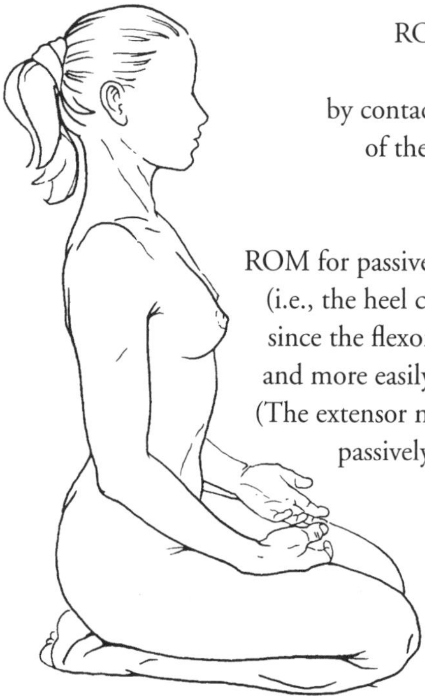

In addition, ROM for flexion is greater
when the hip joint is flexed
and smaller when the hip is extended. Why?

Because position at the hip joint
affects the degree of tension
in the rectus femoris muscle
(see p. 240).

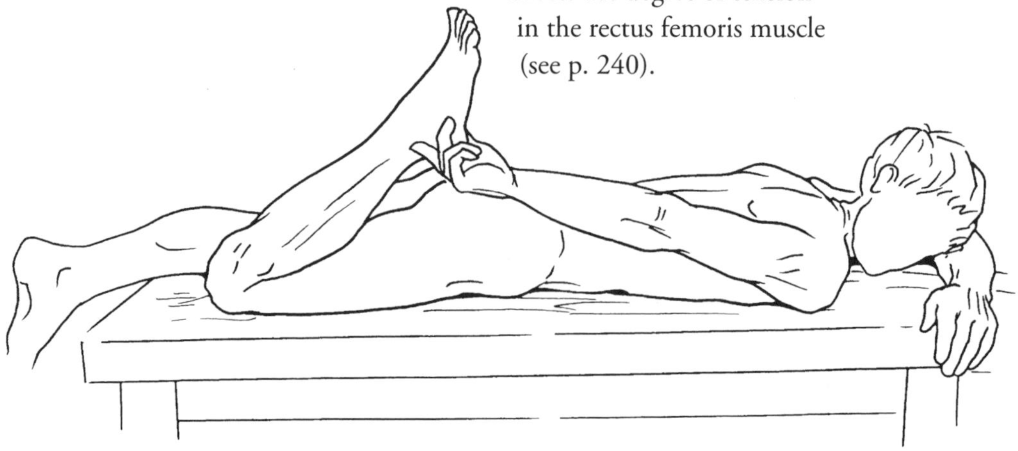

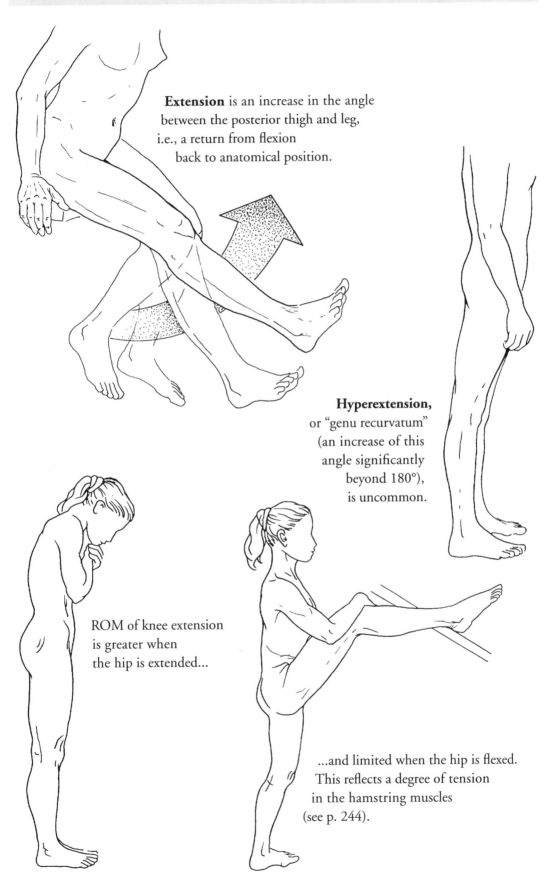

Extension is an increase in the angle between the posterior thigh and leg, i.e., a return from flexion back to anatomical position.

Hyperextension, or "genu recurvatum" (an increase of this angle significantly beyond 180°), is uncommon.

ROM of knee extension is greater when the hip is extended...

...and limited when the hip is flexed. This reflects a degree of tension in the hamstring muscles (see p. 244).

To describe rotation, we will assume that the femur is fixed.

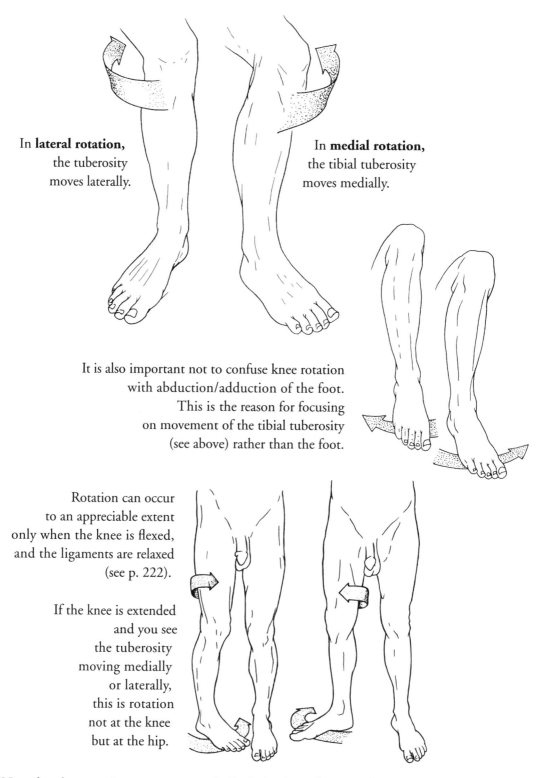

In **lateral rotation,**
the tuberosity
moves laterally.

In **medial rotation,**
the tibial tuberosity
moves medially.

It is also important not to confuse knee rotation
with abduction/adduction of the foot.
This is the reason for focusing
on movement of the tibial tuberosity
(see above) rather than the foot.

Rotation can occur
to an appreciable extent
only when the knee is flexed,
and the ligaments are relaxed
(see p. 222).

If the knee is extended
and you see
the tuberosity
moving medially
or laterally,
this is rotation
not at the knee
but at the hip.

Note that these rotations occur automatically during knee flexion and extension, although ROM is small then and involves both bones (not just the mobile tibia below the femur, as shown above). These rotations are primarily due to the shape of the articular surfaces (see p. 223).

The knee joint consists of three bones

The femur articulates
with the patella,
which is called
the femoropatellar joint.

The femur articulates
with the tibia,
which is called
the femorotibial joint.

**The base
of the femur:**

The shaft of the
femur is triangular
in cross section
(see p. 200).

The patella does
not articulate with
the tibia. We will
study it in detail
on page 225.
Here, we will just
take a look at the
femorotibial joint.

**The top
of the tibia:**

At the bottom:
the posterior edge
of the femur's distal end
bifurcates such that its shape
in cross section changes
to a square, which expands:
thus, the base of the femur
looks like the trunk of a pyramid.

The shaft
of the tibia
is triangular
in cross section.

At the top:
the anterior edge
of the tibia's proximal end
also bifurcates and changes
to an expanding square shape.
Its top looks like
an upside-down pyramid.

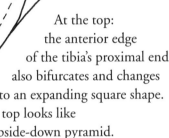

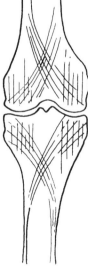

Thus, the two bones
are both expanded
where they come together
and form a massive structure,
like the ends of two columns.
This increases their stability
and weight-bearing ability.

The fibers of the
alveolar (spongy)
tissue inside are
oriented diagonally
and vertically,
as well as horizontally,
which increases
their strength.

Surfaces of knee joint

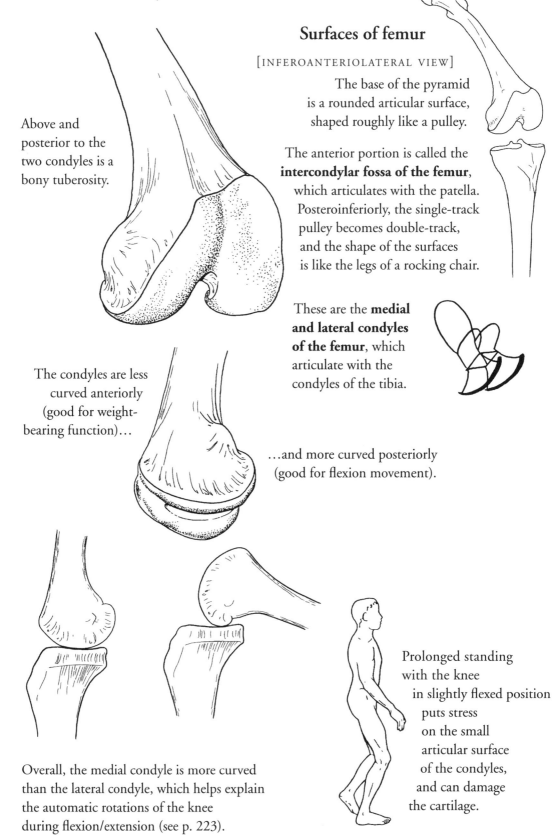

Surfaces of femur

[INFEROANTERIOLATERAL VIEW]

The base of the pyramid is a rounded articular surface, shaped roughly like a pulley.

The anterior portion is called the **intercondylar fossa of the femur**, which articulates with the patella. Posteroinferiorly, the single-track pulley becomes double-track, and the shape of the surfaces is like the legs of a rocking chair.

These are the **medial and lateral condyles of the femur**, which articulate with the condyles of the tibia.

Above and posterior to the two condyles is a bony tuberosity.

The condyles are less curved anteriorly (good for weight-bearing function)…

…and more curved posteriorly (good for flexion movement).

Prolonged standing with the knee in slightly flexed position puts stress on the small articular surface of the condyles, and can damage the cartilage.

Overall, the medial condyle is more curved than the lateral condyle, which helps explain the automatic rotations of the knee during flexion/extension (see p. 223).

Surfaces of tibia

The superior surface (base of the pyramid) of the tibia is called the **tibial plateau**.

The lateral and medial **condyles of the tibia**, protected by cartilage, are concave for articulation with the convex condyles of the femur.

[ANTEROLATERO-SUPERIOR VIEW]

The lateral surface of the tibial plateau has a tubercle, called **Gerdy's tubercle**, where the fascia lata inserts.

At the center of the tibial plateau, the edges of the condyles are raised and form the **intercondylar eminence**.

Its anterior surface has a prominence, the **anterior tibial tuberosity**, where the quadriceps muscle inserts. You can feel this area when you kneel.

Anterior and posterior to the intercondylar eminence are two hollow surfaces, which do not articulate with the femur:

posterior inter-condylar area

anterior inter-condylar area

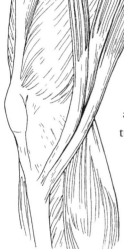

The sartorius, semitendinosus, and rectus femoris muscles attach to the upper medial shaft of the tibia. This is also where the tibial collateral ligament attaches.

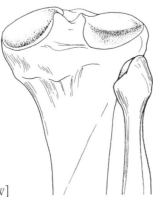

The tibial condyles are concave transversely, but from front to back the lateral condyle is slightly convex while the medial one is concave.

The femorotibial articulation looks like a double-wheel structure fitting into a set of two tracks.

[POSTERIOR VIEW]

Displacement of condyles during movement of knee

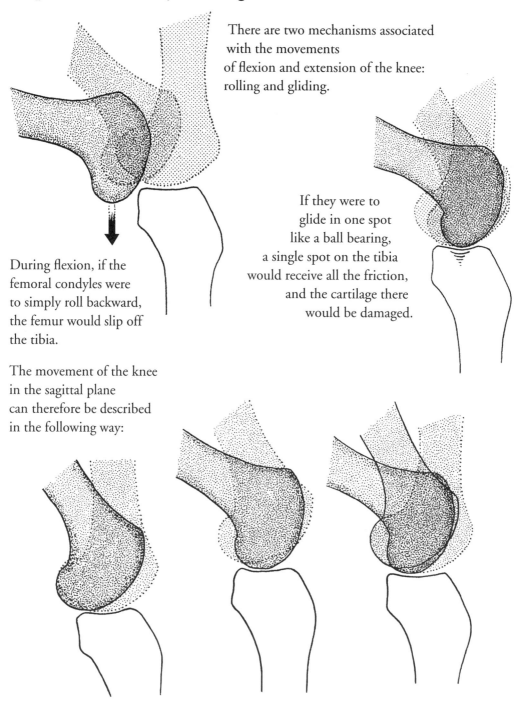

There are two mechanisms associated with the movements of flexion and extension of the knee: rolling and gliding.

During flexion, if the femoral condyles were to simply roll backward, the femur would slip off the tibia.

If they were to glide in one spot like a ball bearing, a single spot on the tibia would receive all the friction, and the cartilage there would be damaged.

The movement of the knee in the sagittal plane can therefore be described in the following way:

In **flexion**, the femoral condyle first rolls (15-20°) on the tibial condyle,

then glides...

producing a combined "rolling-gliding" movement.

The opposite occurs in **extension** of the knee: first gliding, then rolling. During this movement, the lateral condyle rolls more than the medial condyle, which leads to automatic rotation of the knee (see also p. 223).

For the lower limb, in anatomical position, we can consider three different axes.

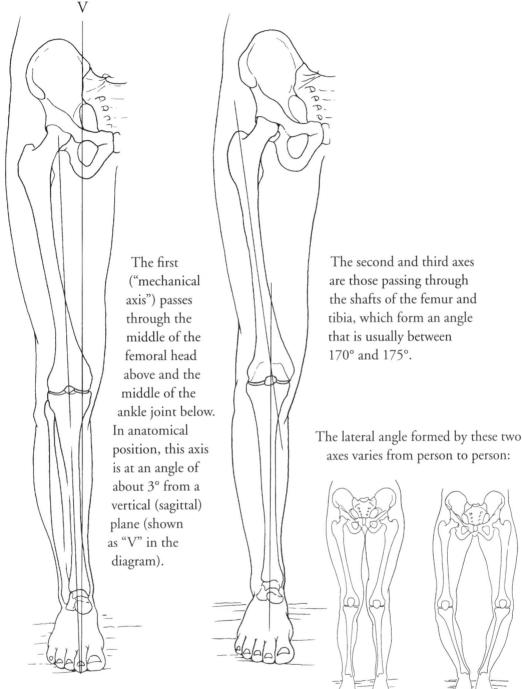

The first ("mechanical axis") passes through the middle of the femoral head above and the middle of the ankle joint below. In anatomical position, this axis is at an angle of about 3° from a vertical (sagittal) plane (shown as "V" in the diagram).

If you stand on one foot, this axis moves farther from the sagittal plane.

The second and third axes are those passing through the shafts of the femur and tibia, which form an angle that is usually between 170° and 175°.

The lateral angle formed by these two axes varies from person to person:

It can be less than 180° (genu valgum or "knock knees"),

...or greater than 180° (genu varum or "bow-legs").

Menisci

The **menisci** (singular: meniscus)
are two croissant-shaped intra-articular discs
made of fibrocartilage.

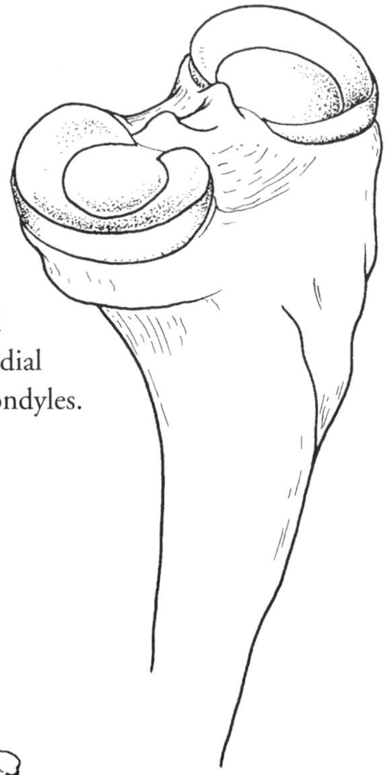

Their shallow central tips are attached
to the intercondylar eminence of the tibia,
and their thicker margins are attached
to the peripheral edges of the medial
and lateral tibial condyles.

They also have attachments
to nearby structures
such as the meniscopatellar ligaments,
medial collateral ligament,
and tendons of the popliteus
and semimembranosus muscles.

The menisci are slightly mobile,
and aid in spreading the synovial fluid
during movements of the knee.

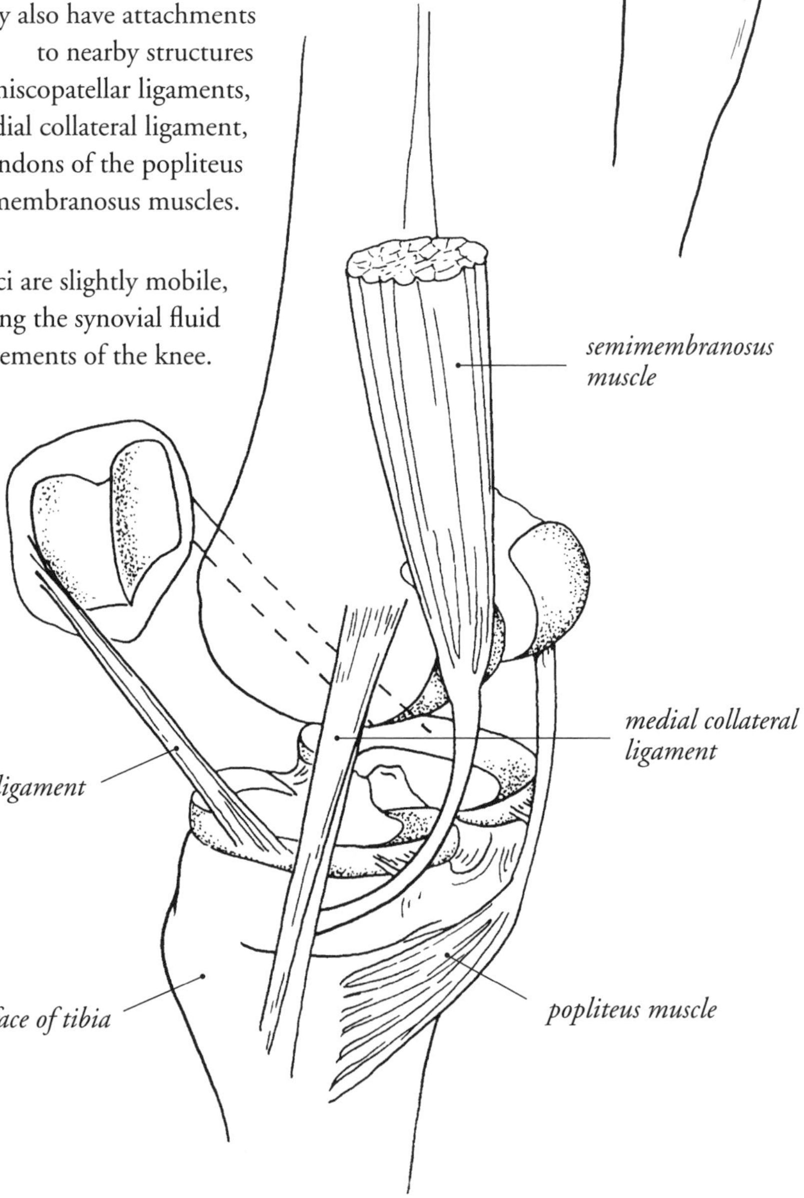

semimembranosus muscle

medial collateral ligament

meniscopatellar ligament

medial surface of tibia

popliteus muscle

[RIGHT KNEE, POSTEROMEDIAL VIEW]

The menisci have several functions:

• As they move around, they increase the distribution of synovial fluid.

• They increase the weight-bearing surface, which results in a better distribution of pressure as they move around.

• Like wedges, they increase the concave shape of the tibial condyles and therefore provide better stability.

without menisci

with menisci

How the menisci move during knee movements

In **extension**, the menisci move forward because they are (1) pushed in that direction by the femoral condyles, and (2) pulled by the meniscopatellar ligaments, which are in turn pulled upward by movement of the patella.

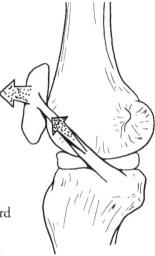

In **flexion**, the menisci move backward because they are (1) pushed in that direction by the condyles…

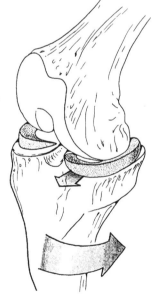

…and (2) pulled by the tendons of the semimembranosus and popliteus muscles, and the medial collateral ligament.

In **rotation**, the ipsilateral meniscus moves forward because of pressure from the condyle and is held back by the meniscopatellar ligament.

These movements are all necessary for normal function of the knee joint. In some cases (particularly rapid extension movements, as in soccer),

…the menisci may not move fast enough, and become crushed or torn.

Knee capsule

The knee joint is held by a thick **capsule,** which attaches just outside the articular surfaces of the three bones involved. The patella is contained in the anterior capsule. Thus the patella, femur, tibia, and capsule enclose a single synovial cavity within which synovial fluid circulates.

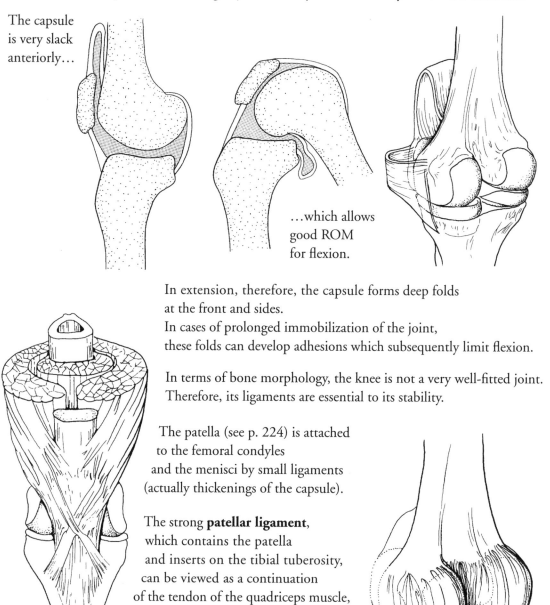

The capsule is very slack anteriorly...

...which allows good ROM for flexion.

In extension, therefore, the capsule forms deep folds at the front and sides.
In cases of prolonged immobilization of the joint, these folds can develop adhesions which subsequently limit flexion.

In terms of bone morphology, the knee is not a very well-fitted joint. Therefore, its ligaments are essential to its stability.

The patella (see p. 224) is attached to the femoral condyles and the menisci by small ligaments (actually thickenings of the capsule).

The strong **patellar ligament,** which contains the patella and inserts on the tibial tuberosity, can be viewed as a continuation of the tendon of the quadriceps muscle, whose fibers cross over each other at the knee joint.

Posteriorly, the knee capsule is thicker and forms two strong bands connecting the femoral and tibial condyles. These resist hyperextension of the joint, and provide "passive stability" in the standing position (see p. 222).

The joint is also held in place by two **cruciate** ("crossed") **ligaments** located in the intercondylar fossa of the femur.

They are named according to where they attach to the tibia. Anatomically, they are outside the joint capsule.

The **anterior cruciate ligament** is attached to the anterior intercondylar area of the tibia. It runs posterosuperolaterally and attaches to the medial aspect of the lateral femoral condyle.

The **posterior cruciate ligament** attaches to the posterior intercondylar area of the tibia. It runs anterosuperomedially and attaches to the lateral surface of the medial femoral condyle.

Their principal role is to prevent anteroposterior displacements of the two bones.

anterior cruciate lig.

posterior cruciate lig.

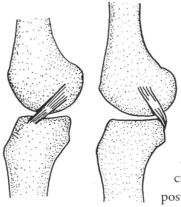

The anterior cruciate ligament tends to resist anterior displacement of the tibia on the femur...

...while the posterior cruciate ligament resists posterior displacement.

Why have obliquely-oriented ligaments perform this braking action?

Because simple anterior and posterior ligaments would not allow flexion.

In both flexion and extension, the cruciate ligaments remain fairly taut, and displacement of the tibia on the femur is minimal.

In lateral rotation, the cruciates slacken somewhat.

In medial rotation, they press against each other, and therefore become more taut.

On the sides, the joint capsule is reinforced by two **collateral ligaments**.

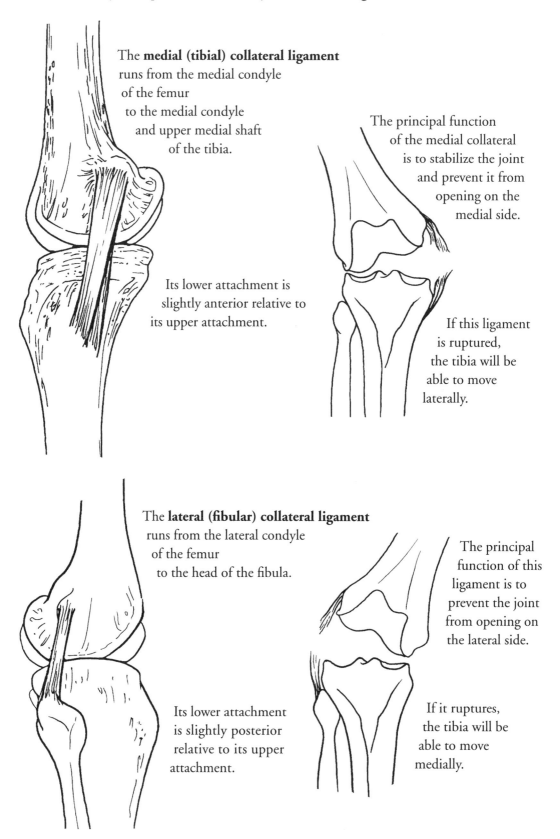

The **medial (tibial) collateral ligament** runs from the medial condyle of the femur to the medial condyle and upper medial shaft of the tibia.

The principal function of the medial collateral is to stabilize the joint and prevent it from opening on the medial side.

Its lower attachment is slightly anterior relative to its upper attachment.

If this ligament is ruptured, the tibia will be able to move laterally.

The **lateral (fibular) collateral ligament** runs from the lateral condyle of the femur to the head of the fibula.

The principal function of this ligament is to prevent the joint from opening on the lateral side.

Its lower attachment is slightly posterior relative to its upper attachment.

If it ruptures, the tibia will be able to move medially.

The medial collateral is considerably thicker and stronger than its lateral counterpart. Why?

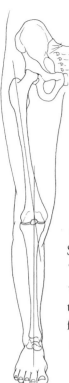

In the average person, the lateral angle formed by the femur and tibia is slightly less than 180° (genu valgum, see p. 215). Since the joint "gapes" more on the medial side, there is a need for stronger stabilization on that side.

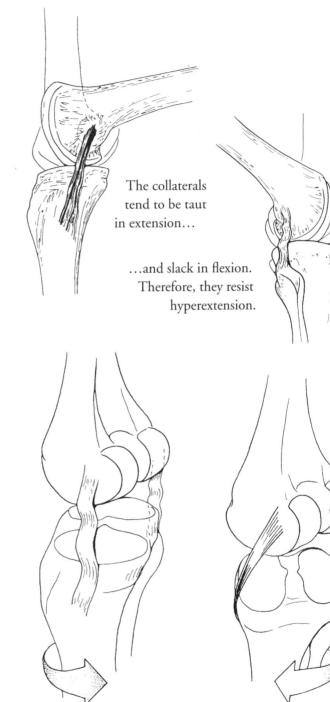

The collaterals tend to be taut in extension...

...and slack in flexion. Therefore, they resist hyperextension.

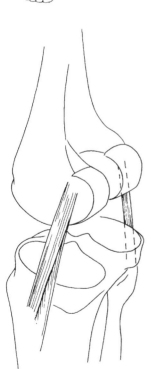

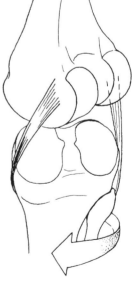

On leg bones, which have been "pulled apart" for illustrative purposes...

we see that the collaterals become slack in medial rotation of the tibia due to their orientation...

...and taut in lateral rotation. Thus, they resist excessive lateral rotation of the tibia.

How ligaments stabilize the knee

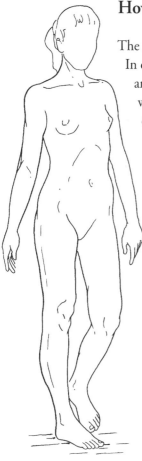

The knee ligaments act together to stabilize the joint.
In **extension**, all the ligaments are taut,
and the joint can be passively stabilized
without any muscular action,
e.g., when balancing on one foot.

Here, the joint is "locked"
in slight hyperextension
by the tautness of the ligaments
(particularly the thickened
posterior portion
of the joint capsule).

In **flexion**,
most of the ligaments are slack,
and the joint therefore
has some capacity for rotation.

As noted above,
the collaterals and cruciates
tend to limit lateral…

…and medial
rotation,
respectively.

But they are more restrictive
in the extended
than in the flexed position.

While balancing on one foot
with the knee flexed,
several muscles are used
to stabilize the body:

• quadriceps to prevent the knee
from flexing more

• rotator muscles to prevent
or slow down excessive rotation

– medially: vastus medialis,
sartorius, gracilis,
semitendinosus

– laterally: vastus medialis,
biceps femoris, tensor fascia
latae (see also muscle actions
on p. 254)

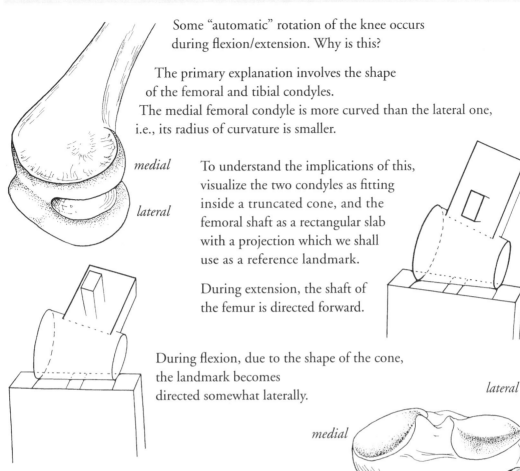

Some "automatic" rotation of the knee occurs during flexion/extension. Why is this?

The primary explanation involves the shape of the femoral and tibial condyles.
The medial femoral condyle is more curved than the lateral one, i.e., its radius of curvature is smaller.

medial

lateral

To understand the implications of this, visualize the two condyles as fitting inside a truncated cone, and the femoral shaft as a rectangular slab with a projection which we shall use as a reference landmark.

During extension, the shaft of the femur is directed forward.

During flexion, due to the shape of the cone, the landmark becomes directed somewhat laterally.

lateral

medial

The tibial condyles are also not totally symmetric; both are concave transversely, but from front to back the lateral condyle is slightly convex while the medial one is concave.
Therefore, the lateral tibial condyle allows more rolling than does the medial one.

[POSTERIOR VIEW]

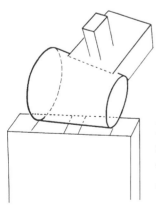

During flexion, the lateral femoral condyle rolls backward more than the medial one does, which accentuates the lateral orientation of our landmark, i.e., the lateral rotation of the femur.

The secondary explanation for automatic rotation of the knee is that the medial collateral ligament is stronger than the lateral one (see p. 221).
This reinforces the tendency of the medial femoral condyle to be less mobile than the lateral one.

Patella

This is a sesamoid bone, located anterior
to the distal end of the femur, which develops
within the tendon of the quadriceps muscle (right).

base

Its anterior surface
sits directly beneath
the skin and is easily
palpable.

apex

The two articular facets on its posterior surface
fit against the patellar surface of the femur,
and are separated by a vertical ridge.

The patella is both attached
to the knee
and is mobile on top of it.

It is connected:

• with the femoral condyles
via the lateral and
medial patellar ligaments

• with the menisci
via the meniscopatellar ligaments.

It is connected
with the tendon
of the quadriceps femoris
via the **patellar ligament**.

What exactly does the patella do? Its primary function is to protect the quadriceps tendon, in which it is contained. During movement, this tendon slides in the groove between the femoral condyles, like a rope in a pulley.

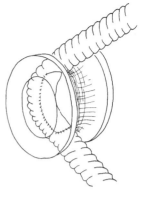

This causes intense strain on the patella:

- strain from being pressed against the groove during flexion due to the pulling action of the quadriceps muscles…

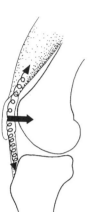

…this pressure can be 400 kg or more during squatting, and even more when carrying a heavy object

- strain from stretching (pulling forces from different directions)

- strain from constant usage.

The patella is not stable laterally.

The quadriceps follows the femoral shaft and its force is slightly oblique, but its tendon runs straight down to insert on the tibia.
Thus, contraction of the quadriceps tends to pull the patella laterally…

…just as a pulley would move sideways if its rope came down at an angle.

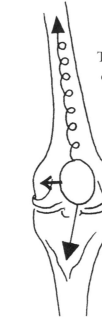

The lateral instability of the patella is maximal
during active extension or slight flexion
when the patella is not firmly pressed
against the patellar surface of the femur;
during full flexion the patella is better "locked" in place.

This instability is accentuated
during lateral rotation of the tibia
when the lower as well as upper part
of the tendon become obliquely oriented.

The tendency of the patella to move laterally
is counteracted by two mechanisms:

• the projection of the lateral femoral
condyle, which is more pronounced
than that of the medial condyle

• the contraction of the vastus medialis muscle,
which pulls the tendon medially.

As you can see, the articulation of the patella
against the femur is subjected to major strains and
stresses, particularly on the lateral side.
This explains the frequency of arthritis here,
which can compromise proper gliding of the patella
and active extension of the knee.

Muscles of the hip and knee with their many bony attachments

KNEE (SHADED)

Sacrum:

superficial fascia of gluteus maximus

Ischium:

semitendinosus
semimembranosus
long head of biceps femoris
gracilis
sartorius
tensor fasciae latae
rectus femoris

Femur:

vastus medialis
vastus lateralis
vastus intermedius
short head of biceps femoris
popliteus

Tibia:

quadriceps muscles
semimembranosus
semitendinosus
gracilis
popliteus
sartorius
tensor fasciae latae
superficial layer of
gluteus maximus

Fibula:

long and short head of biceps
femoris

Patella:

vastus intermedius
vastus medialis
vastus lateralis
rectus femoris

Calcaneus (dotted outline):

gastrocnemius

Femur:

gluteus minimus and medius
gluteus maximus (deep fibers)
adductor muscles (except for
gracilis)
psoas and iliacus
superior and inferior
gemellus muscles
quadratus femoris

HIP

Vertebrae T12/L5:

psoas

Sacrum:

piriformis
gluteus maximus

Ischium:

rectus femoris
sartorius
tensor fasciae latae
gluteus muscles
semitendinosus
semimembranosus
long head of biceps femoris
adductor muscles
obturator muscles
gemellus muscles
quadratus femoris

Coccyx:

gluteus maximus

Patella:

quadriceps muscles

Tibia:

semitendinosus
semimembranosus
gracilis
sartorius
tensor fasciae latae
gluteus maximus (superficial layer)
rectus femoris

Fibula:

long head of
biceps femoris

Muscles of the hip

A group of six deep hip muscles are shown in this inferior view of the pelvis.

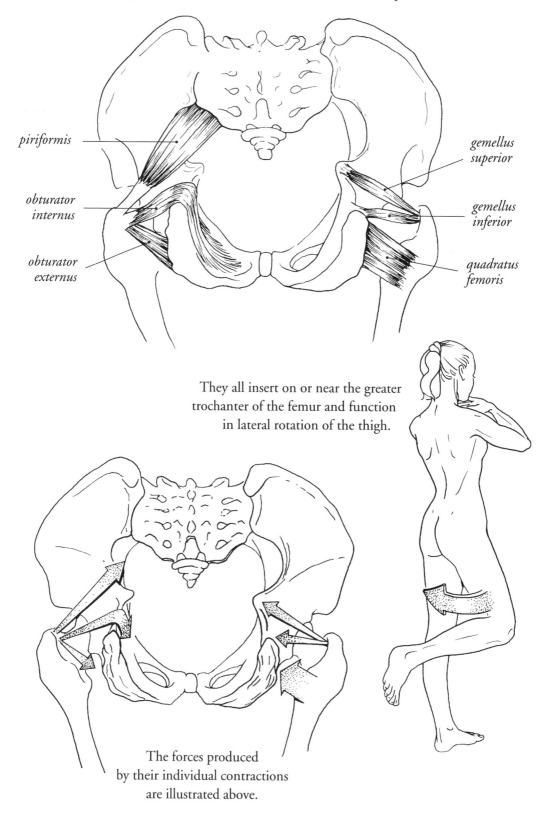

piriformis

obturator internus

obturator externus

gemellus superior

gemellus inferior

quadratus femoris

They all insert on or near the greater trochanter of the femur and function in lateral rotation of the thigh.

The forces produced by their individual contractions are illustrated above.

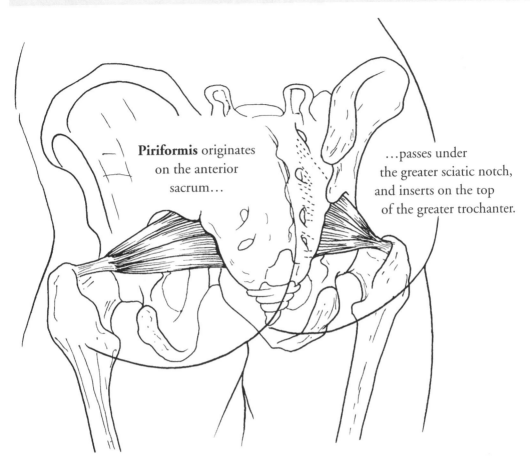

Piriformis originates on the anterior sacrum…

…passes under the greater sciatic notch, and inserts on the top of the greater trochanter.

Actions: If the sacrum is fixed, piriformis laterally rotates, abducts, and flexes the femur.

If the femur is fixed, it contributes to extension of the pelvis (bilateral contraction)…

…or to medial rotation of the pelvis (unilateral contraction).

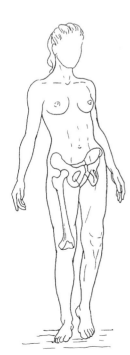

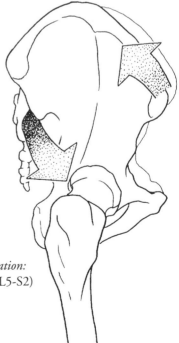

Innervation: sacral plexus (L5-S2)

Quadratus femoris
runs from the lateral ischium,
posterior to the obturator foramen,
and horizontally to the posterior aspect
of the greater trochanter.

Actions: If the pelvis is fixed,
quadratus femoris
laterally rotates the thigh.

If the femur is fixed,
it contributes to extension
of the pelvis (bilateral contraction),
or to medial rotation of the pelvis
(unilateral contraction).

Innervation:
inferior gluteal nerve,
sacral plexus (L5-S2)

The following four muscles insert into the medial surface of the greater trochanter, at the level of a depression called the **trochanteric fossa**.

Obturator internus
arises from the obturator membrane and adjacent portions of the ischium and ilium.

Its fibers pass posteriorly through the lesser sciatic notch, make a sharp bend around the body of the ischium, and insert on the medial aspect of the greater trochanter.

There is a bursa
where it wraps around the ischium,
to reduce friction.

Actions: If the pelvis is fixed,
this muscle laterally rotates, flexes,
and abducts the thigh.

If the femur is fixed,
the muscles act in extension
(bilateral contraction)
and in medial rotation or medial flexion
(unilateral contraction)
of the pelvis.

Innervation:
inferior gluteal nerve,
sacral plexus (L5-S2)

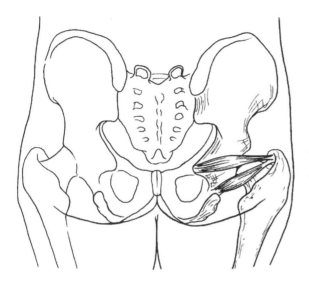

Gemellus superior and inferior
are small muscles located above and below
the distal borders of obturator internus,
at the level of the lesser sciatic notch.
They reinforce the actions
of obturator internus.

Obturator externus
arises from the external surface
of the obturator membrane,
passes posterior to the femoral neck,
and inserts on a fossa on the medial surface
of the greater trochanter.

Actions: If the pelvis is fixed,
it laterally rotates, flexes,
and abducts the femur.

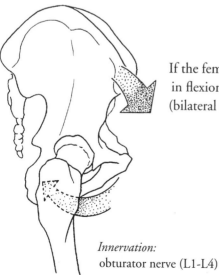

If the femur is fixed, it functions
in flexion of the pelvis
(bilateral contraction)…

…and also medially
rotates and flexes
the pelvis
(unilateral contraction).

Innervation:
obturator nerve (L1-L4)

How obturator and gemellus muscles support the hip

Viewing the hip from the right side, we observe that:

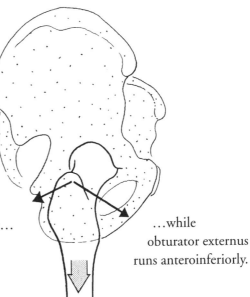

Obturator internus and the gemelli run from the greater trochanter in a posteroinferior direction…

…while obturator externus runs anteroinferiorly.

The combined action of the obturators and gemelli, therefore, can be understood as follows:

• If the pelvis is fixed, they will pull the femur down relative to the pelvis.

• If the femur is fixed, they will lift the pelvis relative to the femur.

Either way, they tend to "pull apart" the hip joint, on a very small scale. This is a decompressive effect which is quite beneficial for certain painful conditions (e.g., worn-down cartilage).

The obturators and gemelli have been compared to a "hammock"…

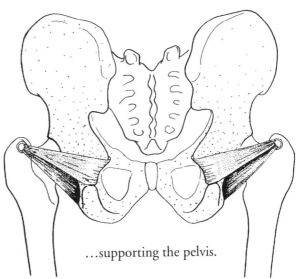

…supporting the pelvis.

Here, the pelvis is tilted backward, showing the two external obturator muscles from the bottom.

In this view, you can see how the muscles wrap under the femoral head and neck before heading superiorlaterally.

Psoas major arises from the bodies
of T12 through L5,
and from arches of fascia
which connect the boney parts
of the vertebral bodies
but do not attach
to the intervertebral disks.

It runs anterior
to the pelvis,
posterior to the
inguinal ligament,
and inserts
on the lesser trochanter.

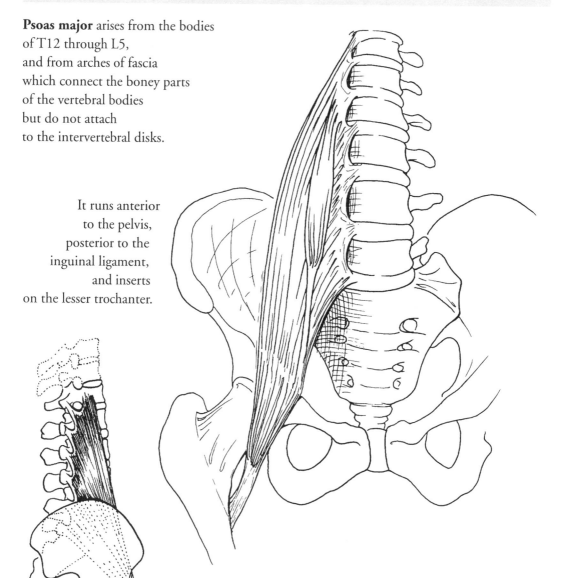

There is a bursa to reduce friction
where it bends at the
anterior pelvis.

Actions: When the vertebrae are
fixed, the psoas flexes the hip,
and works as a weak adductor
and lateral rotator.

Its effect on the lumbar spine
when the femur is fixed was described on page 92.

Innervation: lumbar plexus,
femoral nerve (L1-L3)

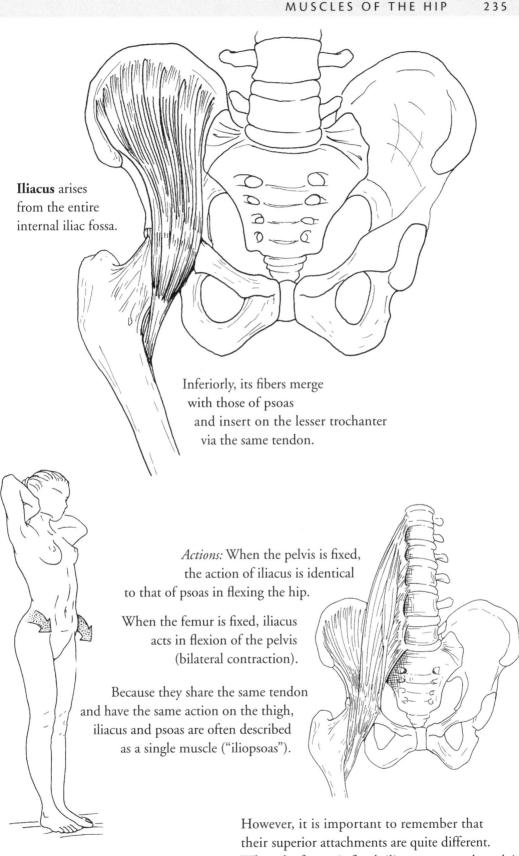

Iliacus arises from the entire internal iliac fossa.

Inferiorly, its fibers merge with those of psoas and insert on the lesser trochanter via the same tendon.

Actions: When the pelvis is fixed, the action of iliacus is identical to that of psoas in flexing the hip.

When the femur is fixed, iliacus acts in flexion of the pelvis (bilateral contraction).

Because they share the same tendon and have the same action on the thigh, iliacus and psoas are often described as a single muscle ("iliopsoas").

Innervation: lumbar plexus, femoral nerve (L2-L4)

However, it is important to remember that their superior attachments are quite different. When the femur is fixed, iliacus acts on the pelvis, whereas psoas acts on the lumbar spine.

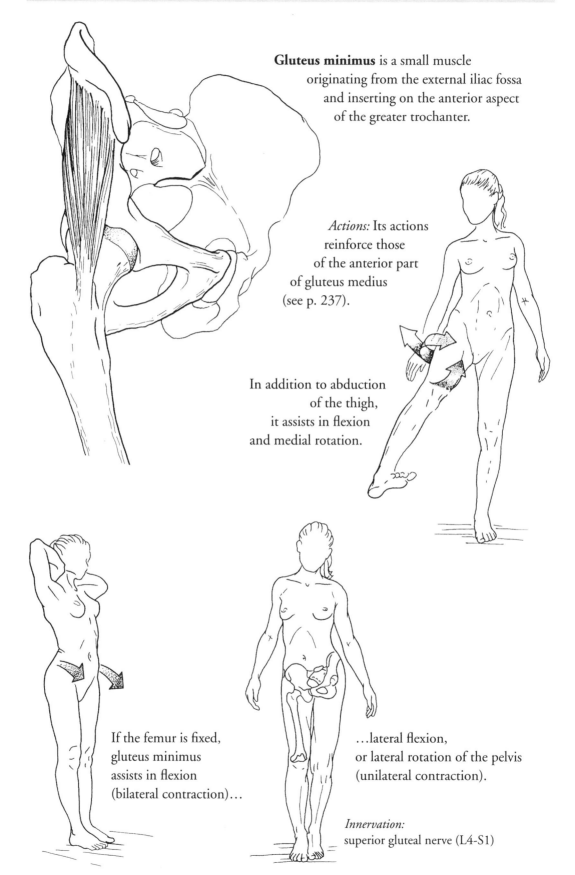

Gluteus minimus is a small muscle originating from the external iliac fossa and inserting on the anterior aspect of the greater trochanter.

Actions: Its actions reinforce those of the anterior part of gluteus medius (see p. 237).

In addition to abduction of the thigh, it assists in flexion and medial rotation.

If the femur is fixed, gluteus minimus assists in flexion (bilateral contraction)…

…lateral flexion, or lateral rotation of the pelvis (unilateral contraction).

Innervation: superior gluteal nerve (L4-S1)

Gluteus medius has a broad origin
on the external iliac fossa.
Its fibers converge and insert
on the lateral aspect
of the greater trochanter.

Actions: When the hip is fixed,
its major action
is abduction of the hip,
but it can also assist in flexion
(anterior fibers)
and extension
(posterior fibers).

When the femur is fixed,
gluteus medius is involved
in both flexion and extension of the pelvis,
depending on whether the anterior or posterior fibers contract
(bilateral contraction).

Its main action is visible when a person stands
on one leg: It mostly acts in lateral flexion of the pelvis.

Innervation:
superior gluteal nerve (L4-L5)

With unilateral contraction, it acts
in lateral flexion of the pelvis, and also
stabilizes the pelvis during walking
(see p. 255) or standing on one foot.

Muscles of the hip and knee

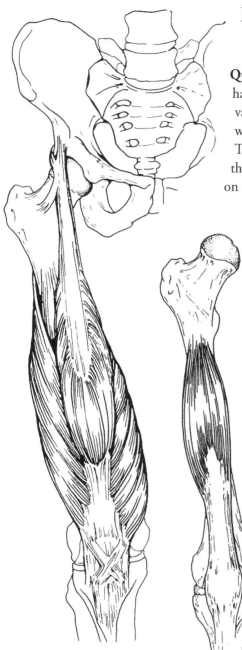

Quadriceps femoris is a massive muscle
having four bodies (rectus femoris, vastus lateralis,
vastus medialis, and vastus intermedius)
which converge into a single quadriceps tendon.
This tendon inserts on and surrounds the patella,
then continues as the patellar ligament to insert
on the tibial tuberosity.

Vastus intermedius, the deepest of the muscle
bodies, originates from the upper two-thirds
of the anterior femoral shaft.
Its fibers follow the axis of the femur.

Vastus intermedius
is covered by
vastus lateralis
and **medialis,** which
arise from either side
of the linea aspera on
the posterior femoral
shaft, then wrap
around the sides
to meet anteriorly,
superficial
to vastus
intermedius.

Rectus femoris arises
from the anterior inferior iliac spine
and part of the ilium near the acetabulum,
and passes superficial to the three vasti
to insert on the common tendon.
Thus, unlike the vasti,
it crosses the hip as well as the knee.

Innervation:
femoral nerve (L2-L4)

This posterior view of the femur shows the origin
of the vastus muscles along the linea aspera (see p. 200).
Vastus medialis has its origin on the medial surface of the femur,
whereas vastus lateralis has its origin on the lateral surface
of the femur. From there, they run on both sides of the femur
toward the anterior thigh.

Actions: With its four
bodies acting together,
the quadriceps extends the knee.
It is the strongest muscle
in the body.

*medial rotation
(vastus medialis)*

*lateral rotation
(vastus lateralis)*

When the knee is flexed,
vastus medialis and lateralis
can play a small part
in rotating the tibia
in their respective directions.

When the knee is extended,
the two muscles act
to stabilize the knee
against lateral displacement,
and thereby actively complement
the actions of the ligaments.

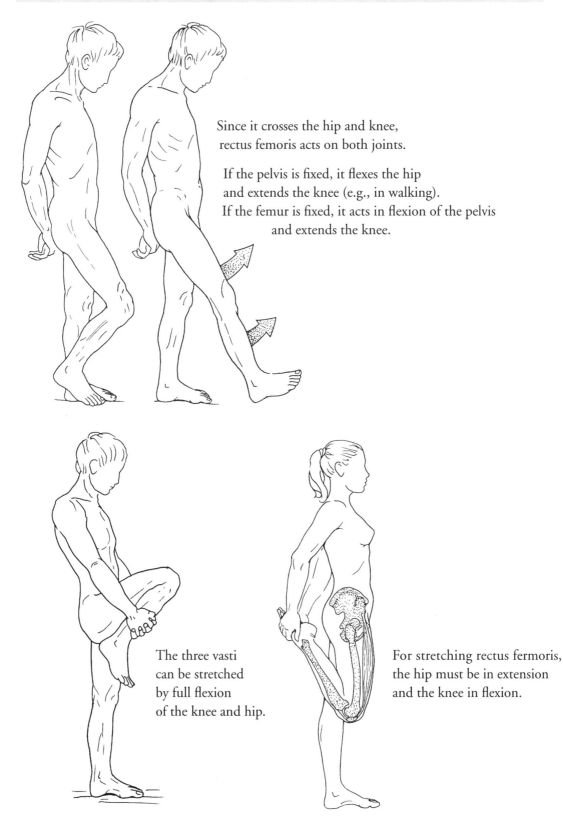

Since it crosses the hip and knee,
rectus femoris acts on both joints.

If the pelvis is fixed, it flexes the hip
and extends the knee (e.g., in walking).
If the femur is fixed, it acts in flexion of the pelvis
and extends the knee.

The three vasti
can be stretched
by full flexion
of the knee and hip.

For stretching rectus fermoris,
the hip must be in extension
and the knee in flexion.

When the rectus femoris is shortened, it is often one of the muscles
responsible for a flexed hip position, i.e., flexion of the pelvis.

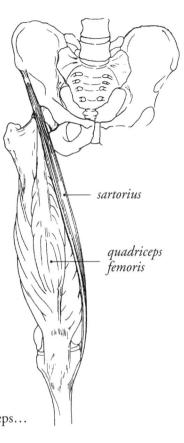

sartorius

quadriceps femoris

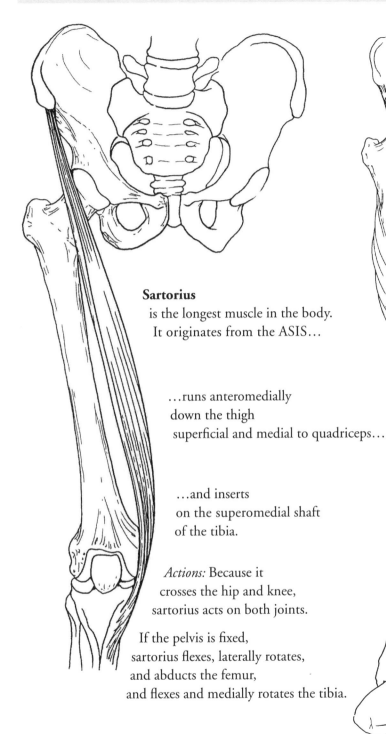

Sartorius
is the longest muscle in the body.
It originates from the ASIS…

…runs anteromedially
down the thigh
superficial and medial to quadriceps…

…and inserts
on the superomedial shaft
of the tibia.

Actions: Because it
crosses the hip and knee,
sartorius acts on both joints.

If the pelvis is fixed,
sartorius flexes, laterally rotates,
and abducts the femur,
and flexes and medially rotates the tibia.

If the leg is fixed, it:
- anteverts the pelvis (bilateral contraction)
- anteverts, medially rotates, and laterally flexes the pelvis
  (unilateral contraction).

Innervation: femoral nerve (L1-L3)

The **hamstrings** are a group of three posterior muscles
working together to flex the knee
and extend the thigh. They all arise
from the ischium, posterior to the hip bone,
and insert on the bones of the lower leg.

Two of these
muscles are
located medially
and insert on the tibia.

*biceps
femoris*

Semitendinosus
is located posterior
to the semimembranosus.
It inserts via a very long,
thin tendon
to the superomedial
tibial shaft.

Semimembranosus
inserts on the
posteromedial aspect
of the tibial condyle.

Biceps femoris is located laterally.
The two heads of this muscle merge inferiorly
and insert via a common tendon to the head of the fibula.
This tendon is bifurcated by the lateral collateral ligament
of the knee.

Innervation:
common tibial
nerve (L5-S2)

Collectively, the hamstrings are polyarticular muscles
of the hip and knee, i.e., they cross and act
on more than one joint.

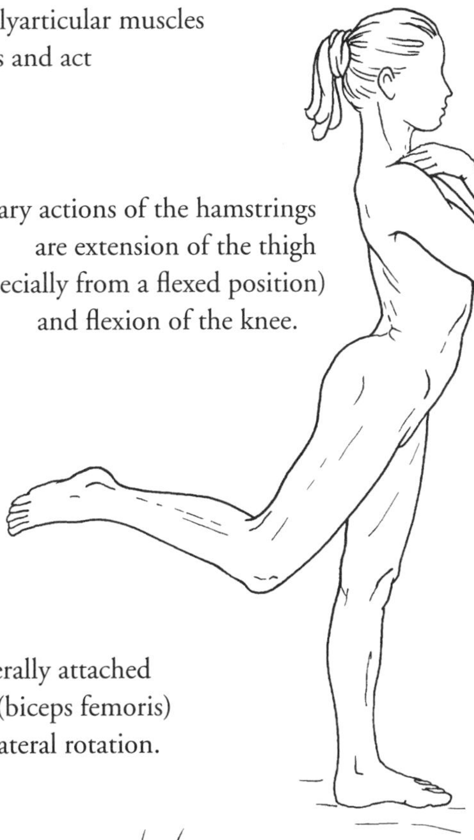

Actions: The primary actions of the hamstrings
are extension of the thigh
(especially from a flexed position)
and flexion of the knee.

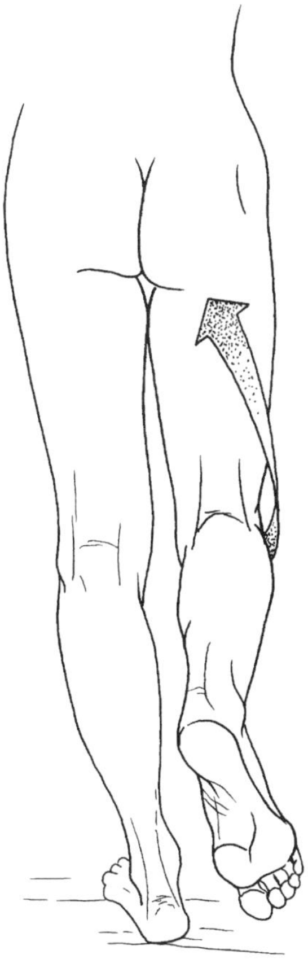

The laterally attached
muscle (biceps femoris)
acts in lateral rotation.

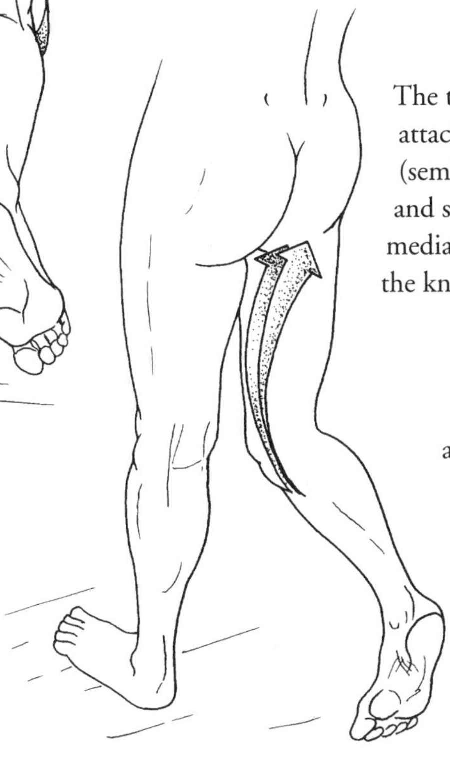

The two medially
attached muscles
(semimembranosus
and semitendinosus)
medially rotate
the knee.

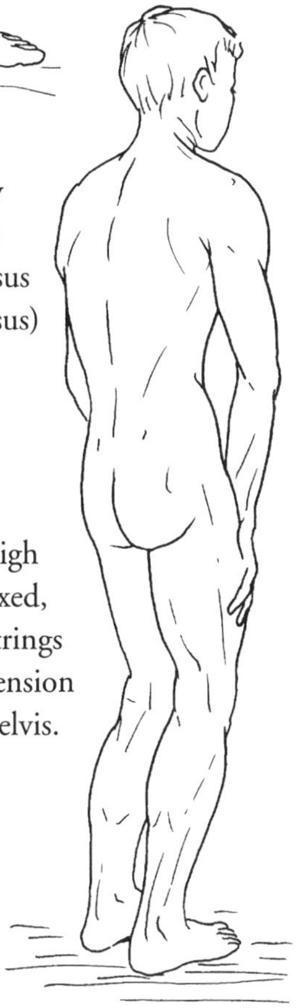

If the thigh
and leg are fixed,
the hamstrings
act in extension
of the pelvis.

In the flexed knee, the tendons of the hamstrings delimit
the **popliteal fossa,** which is readily visible posteriorly.

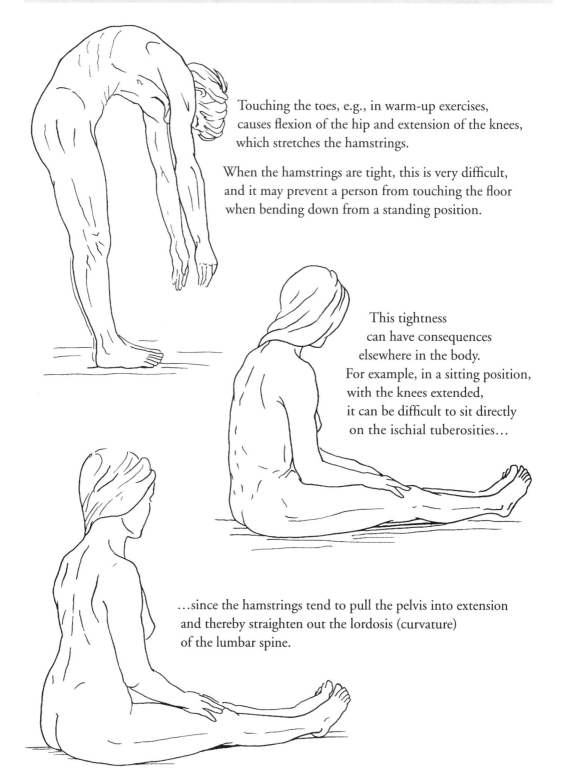

Touching the toes, e.g., in warm-up exercises,
causes flexion of the hip and extension of the knees,
which stretches the hamstrings.

When the hamstrings are tight, this is very difficult,
and it may prevent a person from touching the floor
when bending down from a standing position.

This tightness
can have consequences
elsewhere in the body.
For example, in a sitting position,
with the knees extended,
it can be difficult to sit directly
on the ischial tuberosities…

…since the hamstrings tend to pull the pelvis into extension
and thereby straighten out the lordosis (curvature)
of the lumbar spine.

In this way, tight hamstrings can lead to increased flexion of the lumbar region
and indirectly to disc problems at that level (see p. 42).

It is important to be aware of this problem and do appropriate warm-up exercises on the floor,
especially for beginners.

Adductors

The adductors are a group of five muscles having their bodies on the medial thigh.

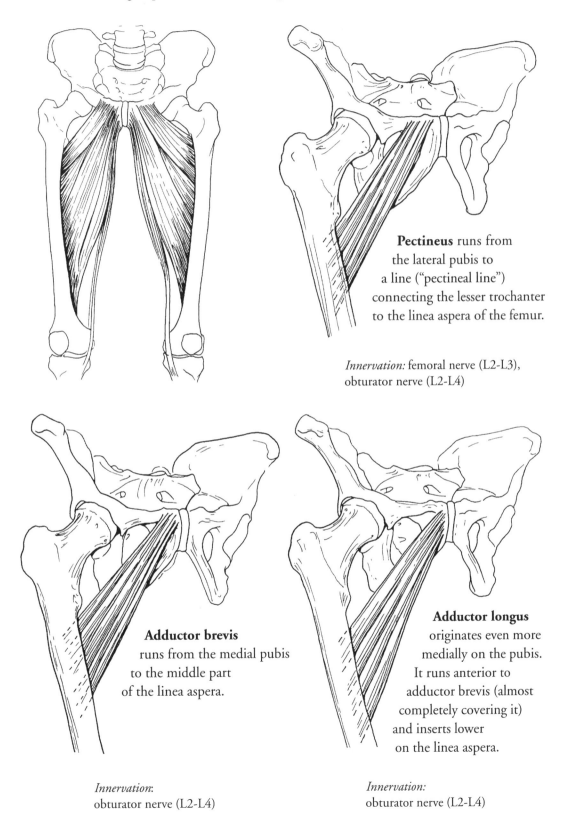

Pectineus runs from the lateral pubis to a line ("pectineal line") connecting the lesser trochanter to the linea aspera of the femur.

Innervation: femoral nerve (L2-L3), obturator nerve (L2-L4)

Adductor brevis runs from the medial pubis to the middle part of the linea aspera.

Innervation: obturator nerve (L2-L4)

Adductor longus originates even more medially on the pubis. It runs anterior to adductor brevis (almost completely covering it) and inserts lower on the linea aspera.

Innervation: obturator nerve (L2-L4)

This illustration shows the two adductor muscles, viewed from the back.
They are easily distinguished from one another.

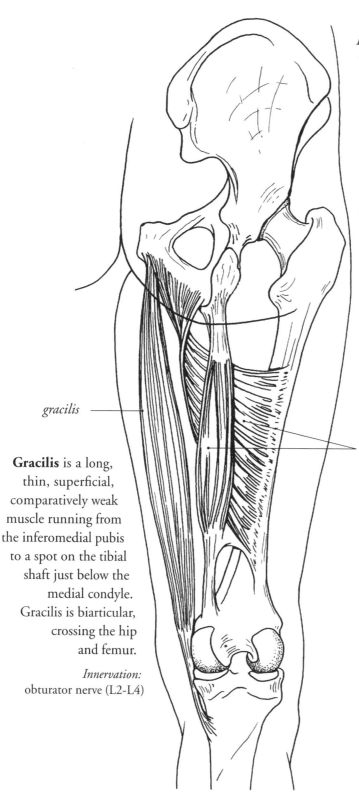

Adductor magnus,
the largest and strongest
of the adductor group,
is really a compound muscle
innervated by two different
spinal nerves.

Its anterior portion originates
from the ischiopubic ramus,
runs inferomedially, and
has a very broad insertion
on the linea aspera.

The posterior portion
originates from the
ischial tuberosity,
runs straight down,
and inserts just above
the medial femoral condyle.

Innervation: obturator nerve,
sciatic nerve (L3-L5)

gracilis

*adductor
magnus*

Gracilis is a long,
thin, superficial,
comparatively weak
muscle running from
the inferomedial pubis
to a spot on the tibial
shaft just below the
medial condyle.
Gracilis is biarticular,
crossing the hip
and femur.

Innervation:
obturator nerve (L2-L4)

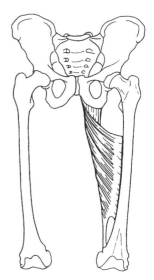

Frontal view of adductor
magnus running
from the ilium to the femur

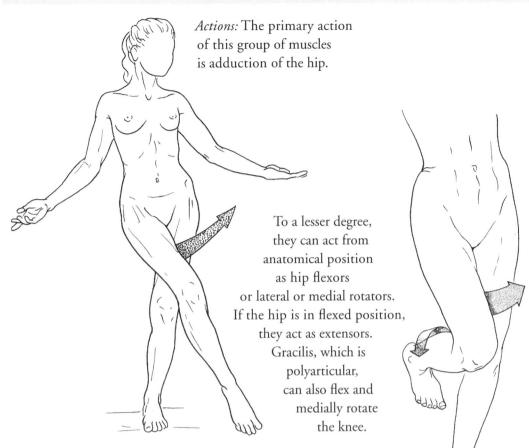

Actions: The primary action of this group of muscles is adduction of the hip.

To a lesser degree, they can act from anatomical position as hip flexors or lateral or medial rotators. If the hip is in flexed position, they act as extensors. Gracilis, which is polyarticular, can also flex and medially rotate the knee.

If the femur is fixed, the adductors are involved in flexion, medial flexion, lateral rotation, or (in the case of gracilis and the posterior portion of adductor magnus) medial rotation of the pelvis.

These muscles, especially gracilis, are frequently strained or torn ("pulled groin") during movements involving sudden or extreme abduction of the thigh.

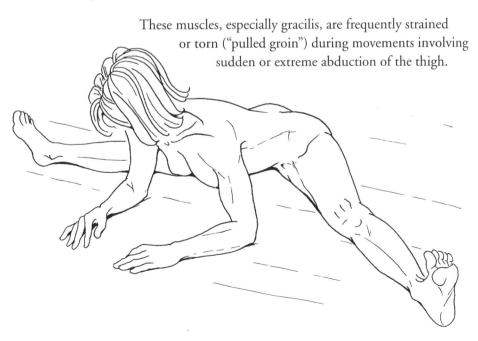

Tensor fasciae latae originates from the anterior iliac crest (near the ASIS) and runs inferiorly and slightly posteriorly.

It inserts not on a bone, but rather on a band of strong fibrous tissue, called the **fascia lata** or iliotibial tract, which runs down the lateral thigh and attaches to the superolateral tibia and head of the fibula.

Actions:
This muscle abducts, flexes, and medially rotates the thigh.

It plays a small part in knee extension or lateral rotation of the flexed knee.

If the thigh and leg are fixed, it acts in flexion (bilateral contraction)…

…lateral flexion, or lateral rotation of the pelvis (unilateral contraction).

Gluteus maximus is one of the largest and strongest muscles of the body. It has two layers: a deep layer and a superficial layer.

It arises on the posterior sacrum and coccyx as well as the posterolateral iliac fossa. The deep layer inserts on the superior linea aspera of the femur, while the superficial layer inserts on the fascia lata.

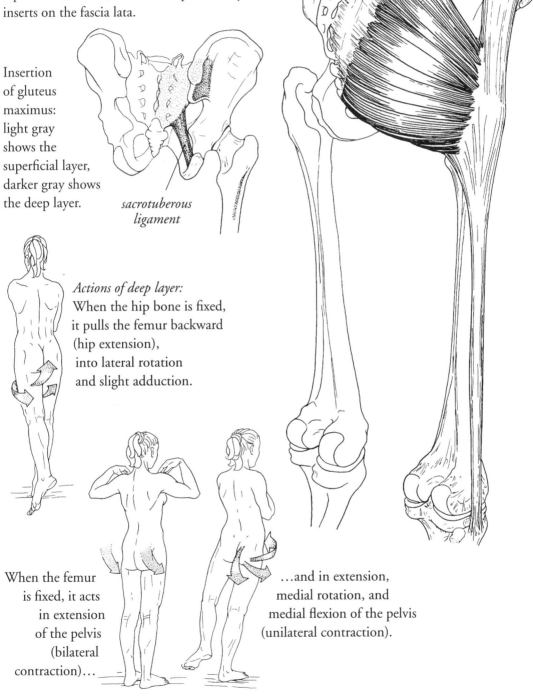

Insertion of gluteus maximus: light gray shows the superficial layer, darker gray shows the deep layer.

sacrotuberous ligament

Actions of deep layer:
When the hip bone is fixed, it pulls the femur backward (hip extension), into lateral rotation and slight adduction.

When the femur is fixed, it acts in extension of the pelvis (bilateral contraction)…

…and in extension, medial rotation, and medial flexion of the pelvis (unilateral contraction).

The action of the superficial layer is discussed together with the tensor fasciae latae on the following page.

The "pelvic deltoid muscle"

The superficial layer of gluteus maximus (in back), and tensor fasciae latae (in front), insert onto the iliotibial tract from opposite directions.

The superficial layer of gluteus maximus acts on the femur by extending, externally rotating, and abducting it.

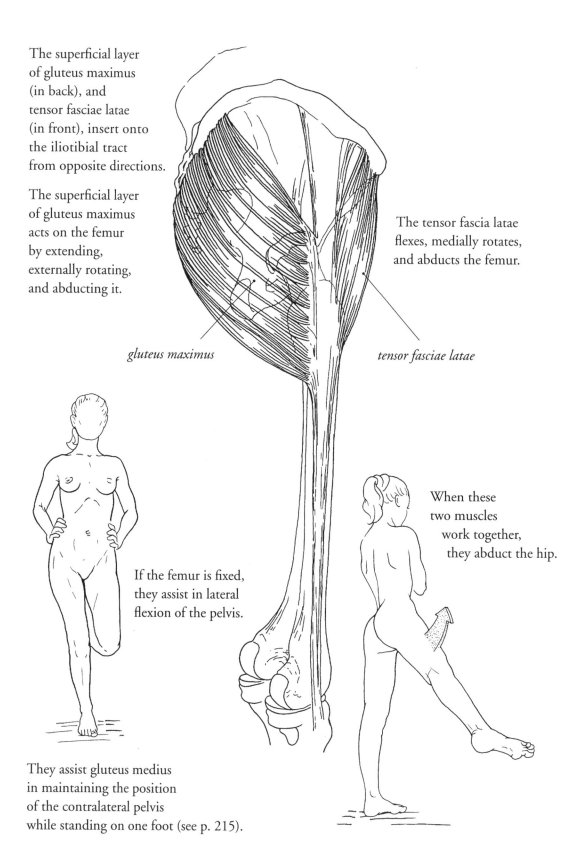

gluteus maximus

tensor fasciae latae

The tensor fascia latae flexes, medially rotates, and abducts the femur.

When these two muscles work together, they abduct the hip.

If the femur is fixed, they assist in lateral flexion of the pelvis.

They assist gluteus medius in maintaining the position of the contralateral pelvis while standing on one foot (see p. 215).

Muscles of the knee

Most muscles acting on the knee also act on the hip, and they have been described above. We need to mention only three muscles which do not cross the hip.

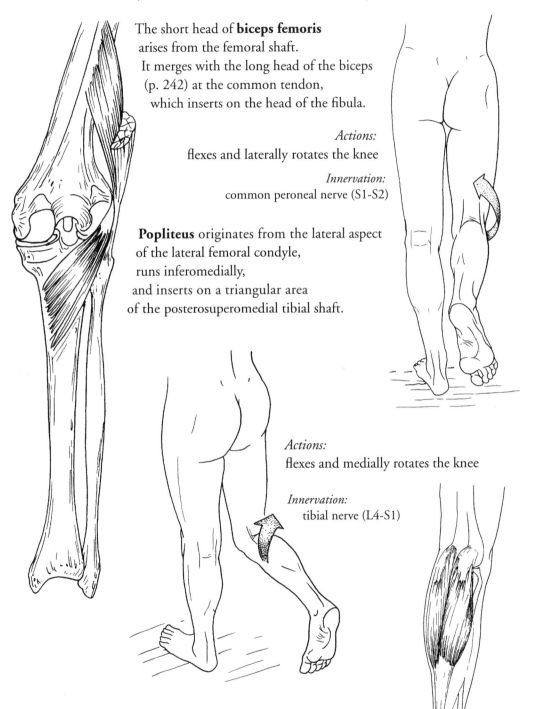

The short head of **biceps femoris** arises from the femoral shaft.
It merges with the long head of the biceps (p. 242) at the common tendon, which inserts on the head of the fibula.

Actions:
flexes and laterally rotates the knee

Innervation:
common peroneal nerve (S1-S2)

Popliteus originates from the lateral aspect of the lateral femoral condyle, runs inferomedially, and inserts on a triangular area of the posterosuperomedial tibial shaft.

Actions:
flexes and medially rotates the knee

Innervation:
tibial nerve (L4-S1)

The **gastrocnemius** are part of the triceps muscles of the calf. They are discussed in detail with the muscles of the ankle on page 292. They flex the knee.

Summary of muscle actions of the hip

The arrows represent the actions produced by the various muscles.

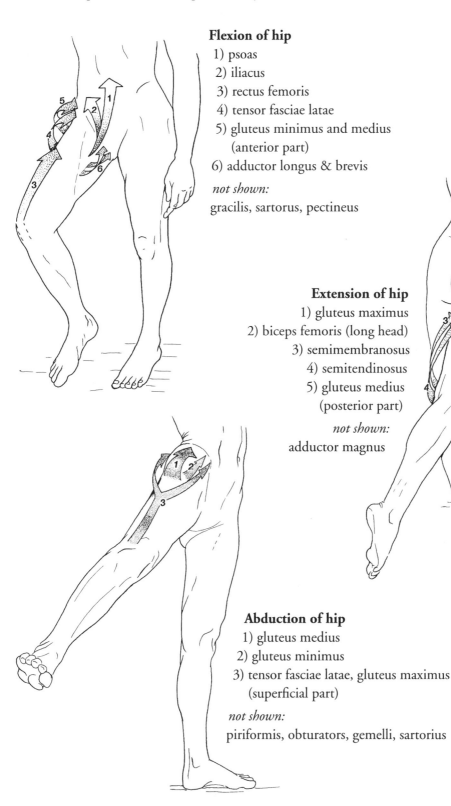

Flexion of hip
1) psoas
2) iliacus
3) rectus femoris
4) tensor fasciae latae
5) gluteus minimus and medius
   (anterior part)
6) adductor longus & brevis

not shown:
gracilis, sartorus, pectineus

Extension of hip
1) gluteus maximus
2) biceps femoris (long head)
3) semimembranosus
4) semitendinosus
5) gluteus medius
   (posterior part)

not shown:
adductor magnus

Abduction of hip
1) gluteus medius
2) gluteus minimus
3) tensor fasciae latae, gluteus maximus
   (superficial part)

not shown:
piriformis, obturators, gemelli, sartorius

Adduction of hip

1) adductor magnus
2) adductor brevis
3) adductor longus
4) pectineus
5) gracilis
6) psoas
7) iliacus

not shown:
biceps femoris (long head),
gluteus maximus (deep part)

Medial rotation of hip

1) gluteus medius
2) gluteus minimus
3) tensor fasciae latae

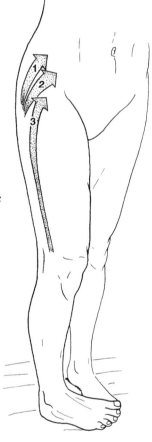

Lateral rotation of hip

1) gluteus maximus

not shown:
piriformis, obturators, gemelli,
quadratus femoris, biceps femoris
(long head), adductors

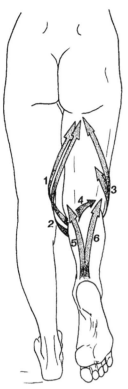

Flexion of knee

1) semitendinosus

2) semimembranosus

3) biceps femoris (long head)

4) popliteus

(5, 6) gastrocnemius (medial & lateral)

not shown:

sartorius, gracilis

Extension of knee

1) quadriceps femoris

2) tensor fasciae latae, gluteus maximus (superficial part)

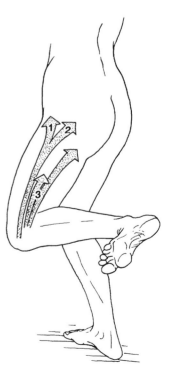

Medial rotation of knee

1) sartorius

2) semitendinosus

3) semimembranosus

4) gracilis

not shown:

popliteus

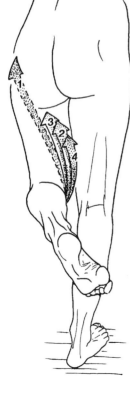

Lateral rotation of knee

1) tensor fasciae latae

2) gluteus maximus (superficial part)

3) biceps femoris (long and short heads)

Actions of muscles of the hip and knee in walking

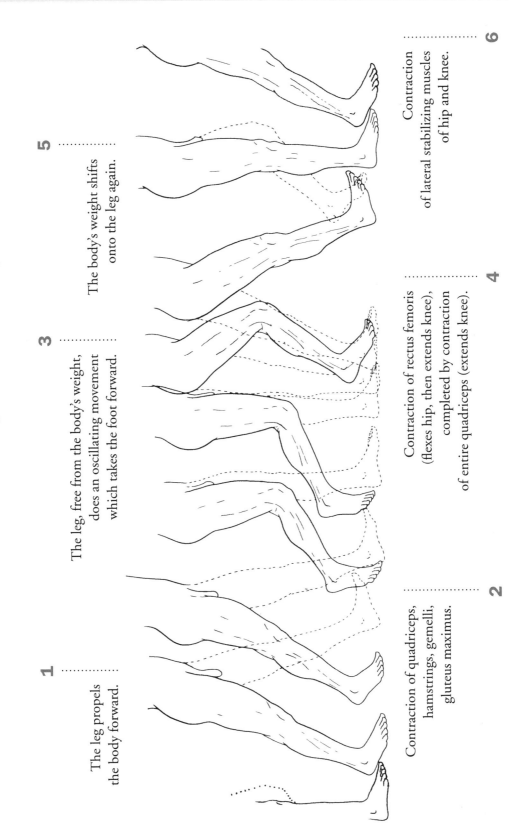

1
The leg propels
the body forward.

2
Contraction of quadriceps,
hamstrings, gemelli,
gluteus maximus.

3
The leg, free from the body's weight,
does an oscillating movement
which takes the foot forward.

4
Contraction of rectus femoris
(flexes hip, then extends knee),
completed by contraction
of entire quadriceps (extends knee).

5
The body's weight shifts
onto the leg again.

6
Contraction
of lateral stabilizing muscles
of hip and knee.

CHAPTER SEVEN

The Ankle & Foot

The **foot**, adapted to a bipedal posture, serves a double function: bearing the weight of the entire body, and performing the dynamic movements necessary for walking. This requires both strength and flexibility.

The foot contains 26 bones, 31 joints, and 20 intrinsic muscles. Unfortunately, malformation of the foot is common due to mechanical stresses from excess weight and poorly-fitted shoes. By understanding its structure and function, we are better able to avoid injury.

The **ankle** joint combines the malleability of the foot with the strength of the leg bones. In this chapter, we will describe both the foot and the ankle joint because the muscles which act on the ankle joint also act on the foot.

Landmarks

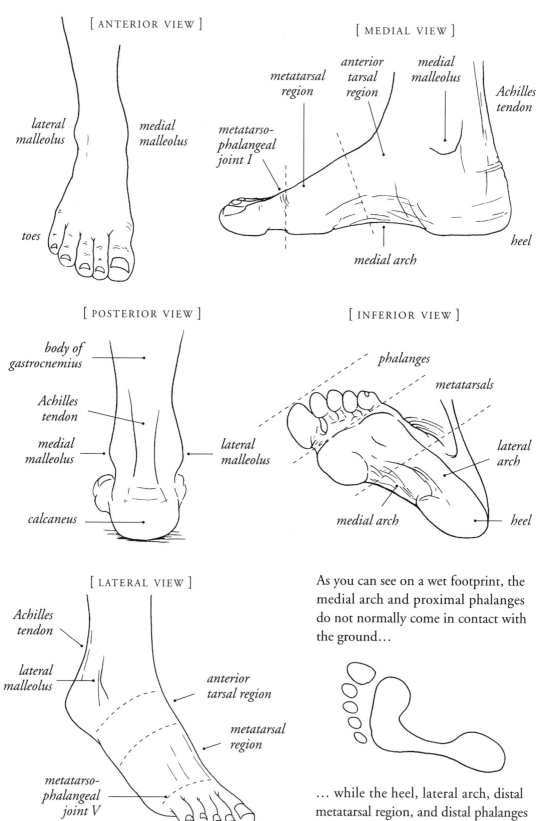

[ANTERIOR VIEW]

lateral malleolus

medial malleolus

toes

[MEDIAL VIEW]

metatarsal region

anterior tarsal region

medial malleolus

Achilles tendon

metatarso-phalangeal joint I

medial arch

heel

[POSTERIOR VIEW]

body of gastrocnemius

Achilles tendon

medial malleolus

lateral malleolus

calcaneus

[INFERIOR VIEW]

phalanges

metatarsals

lateral arch

medial arch

heel

[LATERAL VIEW]

Achilles tendon

lateral malleolus

anterior tarsal region

metatarsal region

metatarso-phalangeal joint V

As you can see on a wet footprint, the medial arch and proximal phalanges do not normally come in contact with the ground...

... while the heel, lateral arch, distal metatarsal region, and distal phalanges do contact it.

Arrangment of bones in foot

A skeletal top view of the foot shows three regions (from front to back):

- In front, there are five slender bones arranged next to each other and flaring outward like "spokes" (numbered I-V from medial to lateral). Each spoke consists of a **metatarsal** which is extended by a **phalange** (toe).

- In back, there are two sizeable bones sitting on top of each other: the **talus** and the **calcaneous**. This part of the foot is known as the **hindfoot** or **posterior tarsus**.

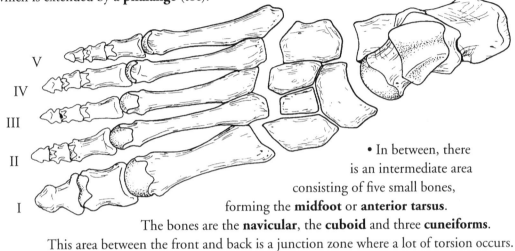

- In between, there is an intermediate area consisting of five small bones, forming the **midfoot** or **anterior tarsus**. The bones are the **navicular**, the **cuboid** and three **cuneiforms**. This area between the front and back is a junction zone where a lot of torsion occurs. This helps the foot adjust to the surface of the ground.

Viewed from medial to lateral, the bony structure of the foot looks fork-shaped, such that:

cuboid   *calcaneus*

- A "lateral foot" follows the calcaneus. It lengthens into two lateral spokes (IV and V). This part of the foot is involved in receiving the body's weight.

cuneiforms

talus

navicular

- A "medial foot" follows the talus. It lengthens into three medial spokes (I-III). This part of the foot is involved in propulsion.

Movements of the foot

We can speak of the following movements in reference to the entire foot, or to some specific region or joint.

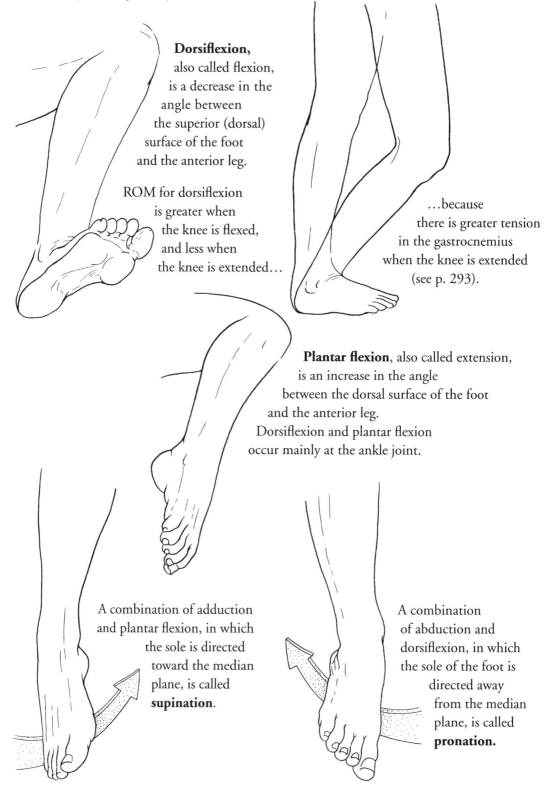

Dorsiflexion, also called flexion, is a decrease in the angle between the superior (dorsal) surface of the foot and the anterior leg.

ROM for dorsiflexion is greater when the knee is flexed, and less when the knee is extended…

…because there is greater tension in the gastrocnemius when the knee is extended (see p. 293).

Plantar flexion, also called extension, is an increase in the angle between the dorsal surface of the foot and the anterior leg. Dorsiflexion and plantar flexion occur mainly at the ankle joint.

A combination of adduction and plantar flexion, in which the sole is directed toward the median plane, is called **supination.**

A combination of abduction and dorsiflexion, in which the sole of the foot is directed away from the median plane, is called **pronation.**

In **abduction** and **adduction,** the distal end of the foot moves away from and toward the median plane, respectively.

These movements can be amplified by or confused with medial and lateral rotation of the hip (when the knee is extended)...

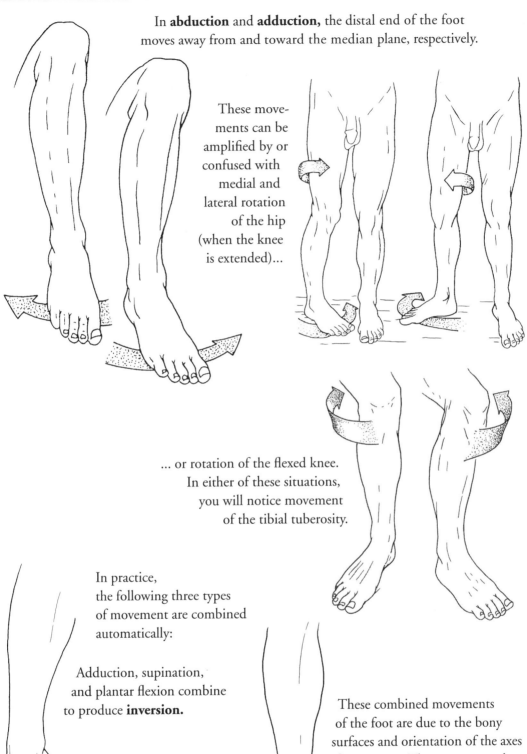

... or rotation of the flexed knee. In either of these situations, you will notice movement of the tibial tuberosity.

In practice, the following three types of movement are combined automatically:

Adduction, supination, and plantar flexion combine to produce **inversion.**

Adbuction, pronation, and dorsiflexion combine to produce **eversion**.

These combined movements of the foot are due to the bony surfaces and orientation of the axes of movement. They are executed simultaneously while walking (see also p. 271).

Two long bones form the skeleton of the leg, tibia, and fibula

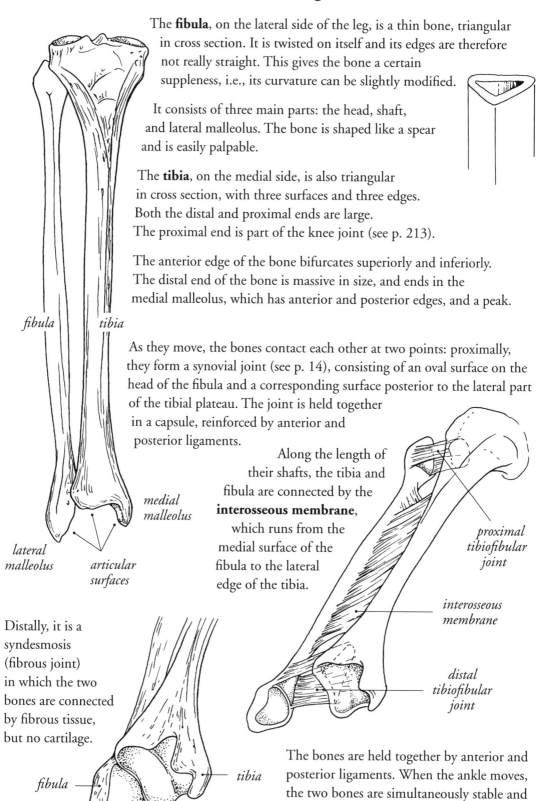

The **fibula**, on the lateral side of the leg, is a thin bone, triangular in cross section. It is twisted on itself and its edges are therefore not really straight. This gives the bone a certain suppleness, i.e., its curvature can be slightly modified.

It consists of three main parts: the head, shaft, and lateral malleolus. The bone is shaped like a spear and is easily palpable.

The **tibia**, on the medial side, is also triangular in cross section, with three surfaces and three edges. Both the distal and proximal ends are large. The proximal end is part of the knee joint (see p. 213).

The anterior edge of the bone bifurcates superiorly and inferiorly. The distal end of the bone is massive in size, and ends in the medial malleolus, which has anterior and posterior edges, and a peak.

As they move, the bones contact each other at two points: proximally, they form a synovial joint (see p. 14), consisting of an oval surface on the head of the fibula and a corresponding surface posterior to the lateral part of the tibial plateau. The joint is held together in a capsule, reinforced by anterior and posterior ligaments.

Along the length of their shafts, the tibia and fibula are connected by the **interosseous membrane**, which runs from the medial surface of the fibula to the lateral edge of the tibia.

fibula   *tibia*

medial malleolus

lateral malleolus   *articular surfaces*

proximal tibiofibular joint

interosseous membrane

distal tibiofibular joint

Distally, it is a syndesmosis (fibrous joint) in which the two bones are connected by fibrous tissue, but no cartilage.

fibula   *tibia*

The bones are held together by anterior and posterior ligaments. When the ankle moves, the two bones are simultaneously stable and mobile. Distally, they "clasp" the talus (the uppermost tarsal bone).

Ankle joint

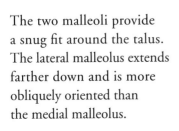

Seen from the front,
this joint resembles
a pincer or crescent wrench
gripping a section
of a hemisphere.

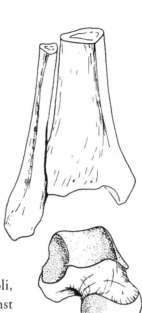

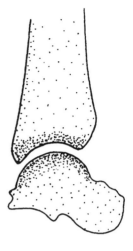

The lateral and medial malleoli,
and distal tibia, fit against
the three facets of the talar body
(see below).

Seen from the side,
the cartilaginous superior
and inferior articulating
surfaces appear
concave and convex,
respectively.

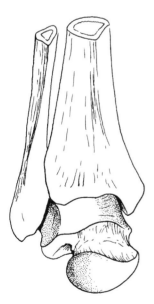

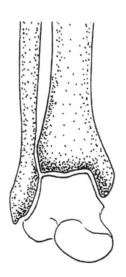

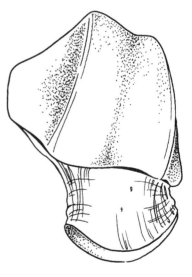

The two malleoli provide
a snug fit around the talus.
The lateral malleolus extends
farther down and is more
obliquely oriented than
the medial malleolus.

In cross section, we see
that there is a slight ridge
on the articulating surface
of the distal tibia, and
a corresponding groove
on talus.

We should also note that
the superior talus is wider
anteriorly than posteriorly.

Mobility of the ankle

Because of the tight fit of the bones, the ankle is a pure hinge joint, capable only of dorsiflexion...

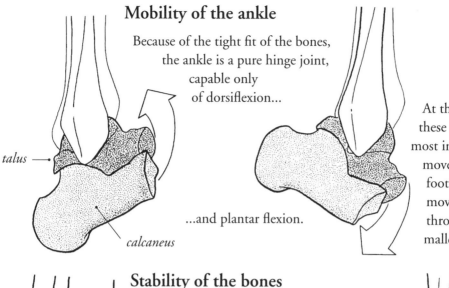

talus

calcaneus

...and plantar flexion.

At this level, these are the most important movements of the foot. The axis of movement passes through the two malleoli.

Stability of the bones

Of all the joints in the foot, this one allows the greatest range of motion (ROM). In dorsiflexion (left), the anterior (wider) part of the superior talus moves into the "pincer," and the joint is more stable.

In plantar flexion (right), the posterior (narrower) part of the talus is in the "pincer," and the joint is therefore less stable. This lack of stability is compensated for in part by support from surrounding muscles and ligaments (see p. 295).

Capsule and ligaments of the ankle

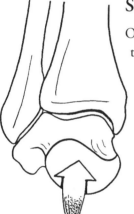

This joint is held together by a capsule which attaches to the proximal joint surfaces of the tibia, fibula, and talus. The capsule is slack anteriorly and posteriorly, allowing for plantar flexion and dorsiflexion of the foot. It is primarily reinforced by lateral ligaments, which are arranged roughly symmetrically: three ligamentous fasciae on each side arise from a malleolus and run obliquely down the two bones of the hindfoot.

talofibular ligaments:

posterior   *anterior*

calcaneofibular ligament

Laterally, the **anterior** and **posterior talofibular ligaments** end on the talus and tie it directly to the bones of the leg, while the **calcaneofibular ligament** runs down to the lateral calcaneus, which is thus involved in ROM of the ankle.

Medially, the **medial collateral ligament** (or **deltoid ligament**) consists of three fasciae in two layers:

• a superficial layer with fibers that insert on the navicular bone, calcaneonavicular ligament, and sustentaculum tali. This layer completely covers the deeper layer.

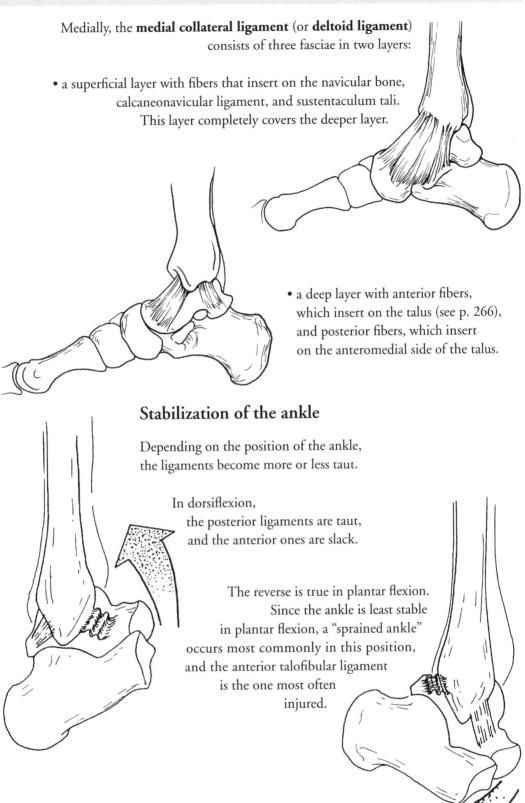

• a deep layer with anterior fibers, which insert on the talus (see p. 266), and posterior fibers, which insert on the anteromedial side of the talus.

Stabilization of the ankle

Depending on the position of the ankle, the ligaments become more or less taut.

In dorsiflexion, the posterior ligaments are taut, and the anterior ones are slack.

The reverse is true in plantar flexion. Since the ankle is least stable in plantar flexion, a "sprained ankle" occurs most commonly in this position, and the anterior talofibular ligament is the one most often injured.

The stability of the ankle is also assisted by muscular action, which helps adjust the "pincer" action, making it more or less tight, during active movement of the ankle (see p. 295).

Talus and calcaneus

These are the most posterior and massive of the tarsal bones.
Calcaneus, at the bottom, articulates with the heel region.
Talus, above, articulates with the ankle (see p. 263).

[MEDIAL VIEW]

talus   calcaneus

These bones are like two rectangular boxes
set at an angle on top of each other,
with the talus pointing medially
and the calcaneous pointing
laterally.

They consist of six surfaces:
superior, inferior, medial,
lateral, anterior,
and posterior.

[LATERAL VIEW]

Talus articulates with the tibia and fibula above,
with calcaneus below, and with navicular in front.
Interestingly, no muscle inserts on talus.
It is moved indirectly via the structures surrounding it.

Calcaneus articulates with talus above
and cuboid in front.

body of
talus

talus

head of talus

calcaneus

trochlear process

On this page and the next, we will take a closer look at these two bones from different angles.

[LATERAL VIEW]

Talus

On its posterosuperior and lateral surfaces is the **trochlea of the talus**. More anteriorly is the **neck of the talus**. And even further anteriorly is the **head of the talus** with a semi-circular articular surface.

It articulates, in turn, with the navicular bone, calcaneonavicular ligament, and the superior surface of the calcaneus (see following pages).

Calcaneus

The anterior surface consists mainly of a triangular articulation, which is concave superiorly and convex inferiorly, and articulates with the posterior surface of the cuboid.

Its inferior surface has two tuberosities (medial and lateral), which contact the ground. At the lateral, more anterior surface of the calcaneus is a bony eminence, the **trochlear process**, which separates the peroneus tendons.

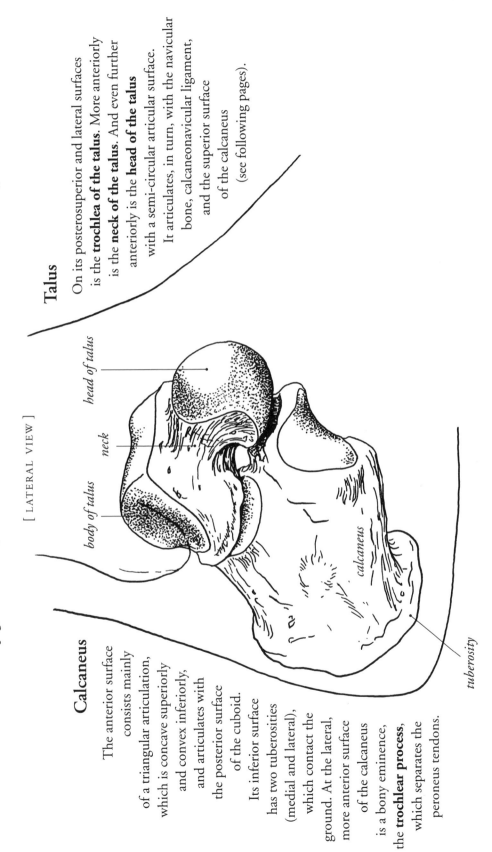

head of talus

neck

body of talus

calcaneus

tuberosity

[MEDIAL VIEW FROM BACK]

The posterior surface of **talus** constitutes the back of its trochlea. Inferiorly, there are two lateral tubercles, which are separated by a groove where the flexor hallucis longus tendon passes.

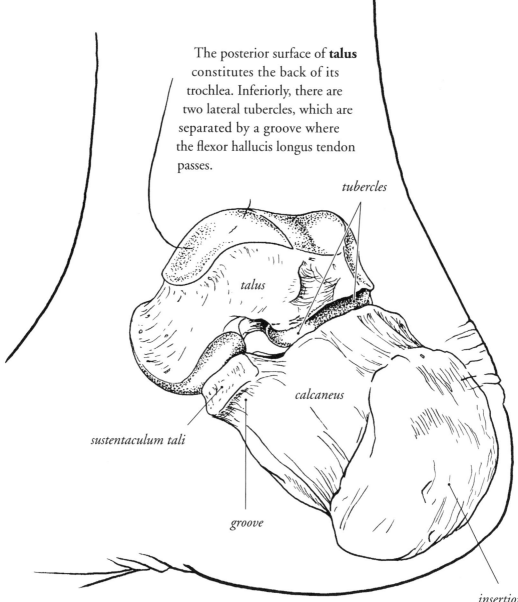

tubercles

talus

calcaneus

sustentaculum tali

groove

insertion area

On **calcaneus**, we see a prominent projection called the **sustentaculum tali** (helps support talus), a groove for passage of various tendons, blood vessels and nerves, and a large insertion area for the Achilles tendon.

Subtalar (talocalcaneal) joint

Talus sits slightly obliquely on calcaneus, since the long axes of the two bones are oriented (respectively) somewhat medially and laterally.

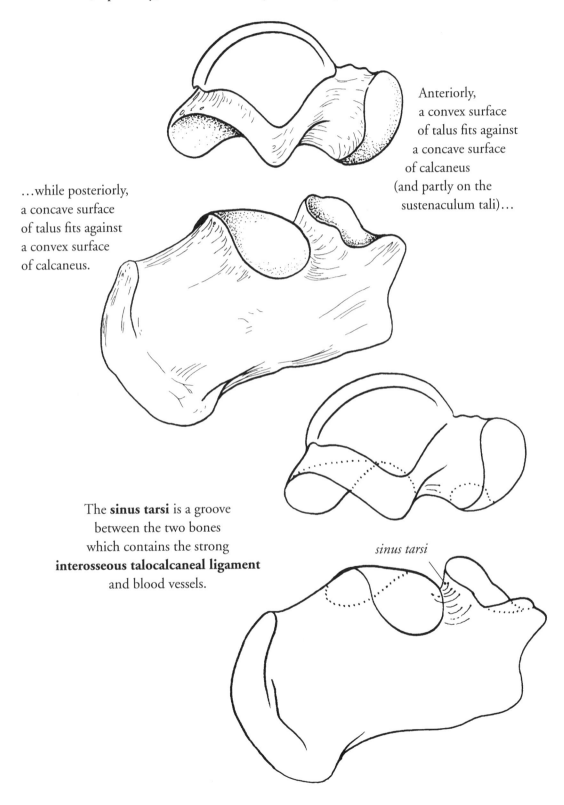

Anteriorly, a convex surface of talus fits against a concave surface of calcaneus (and partly on the sustenaculum tali)…

…while posteriorly, a concave surface of talus fits against a convex surface of calcaneus.

The **sinus tarsi** is a groove between the two bones which contains the strong **interosseous talocalcaneal ligament** and blood vessels.

sinus tarsi

Mobility of subtalar joint

The subtalar joint sits below the ankle, at a vertical angle.
This gives it greater ROM than the ankle, but its ROM is still limited.
We will look at possible movements in three different planes, with and without support.

Frontal plane (back view, with support): calcaneous is tilting sideways below the talus

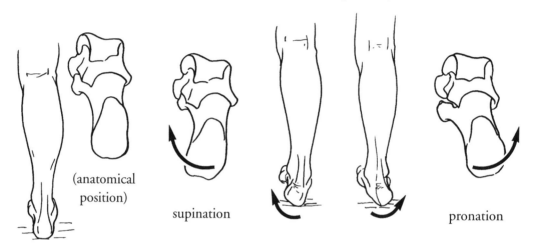

(anatomical position)   supination   pronation

Sagittal plane (without support): calcaneous is moving front to back

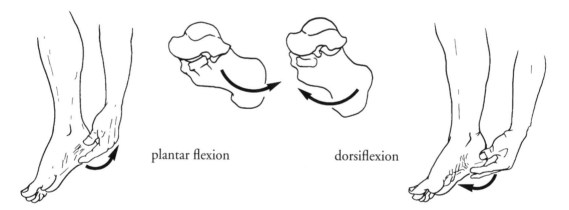

plantar flexion   dorsiflexion

Transverse plane (view from above): calcaneous is moving while turning under the talus

medial   *lateral*

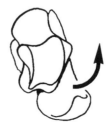

abduction   adduction

Actually, because of the structure of the articulating surfaces, these movements tend
to be combined around an imaginary line called the axis of Henke (a German anatomist).

axis of Henke

This axis enters
the posterolateral tuberosity
of calcaneus,

This axis
therefore
runs obliquely
from back to front,
bottom to top,
lateral to medial.

...runs
anterosuperomedially,

...and exits through
the medial neck of talus.

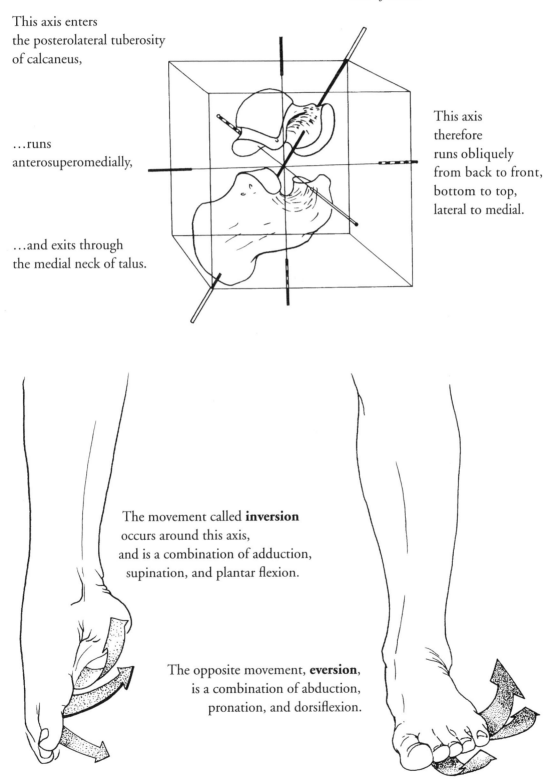

The movement called **inversion**
occurs around this axis,
and is a combination of adduction,
supination, and plantar flexion.

The opposite movement, **eversion**,
is a combination of abduction,
pronation, and dorsiflexion.

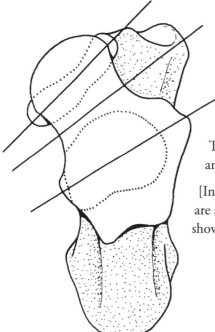

The long axes of the two articulating surfaces
and the sinus tarsi are directed anterolaterally.

[In this see-through view, the two bones
are superimposed, with their articulating surfaces
shown by the dotted outline.]

Capsule and ligaments of subtalar joint

The surfaces are held together by

- two capsules:

  – posteriorly, a capsule which attaches
    to the circumference of the surfaces

  – anteriorly, a capsule shared
    with the transverse tarsal joint.

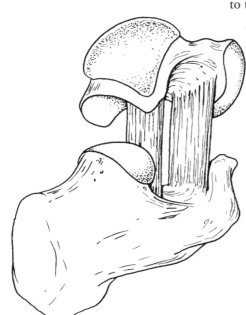

Because the surfaces (on the talus)
and the capsules are continuous,
the movements of the anterior subtalar
and the transverse tarsal joints are inseparable.

- ligaments:

  – The **interosseous talocalcaneal ligament**
    forms a double row of ligaments
    passing along the tarsal tunnel.

  – There is also an anterior
    and a posterior ligament.

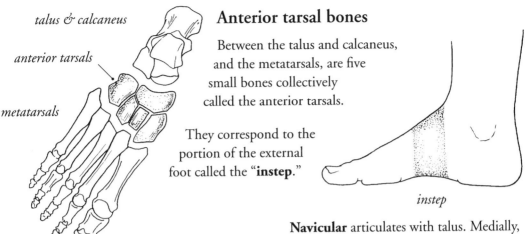

talus & calcaneus

anterior tarsals

metatarsals

Anterior tarsal bones

Between the talus and calcaneus, and the metatarsals, are five small bones collectively called the anterior tarsals.

They correspond to the portion of the external foot called the "**instep**."

instep

Navicular articulates with talus. Medially, it has an externally palpable tubercle for insertion of tibialis posterior (see p. 290).

process   *tubercle*

notch   *navicular*

Cuboid does not actually bear much resemblance to a cube.

cuboid

It has facets for articulation with calcaneus (proximally), lateral cuneiform and navicular (medially), and metatarsals IV and V (distally). A proximal process fits under calcaneus and helps maintain the lateral arch of the foot. A lateral notch continues as a groove on the inferior surface and accommodates the tendon of peroneus longus (see p. 288).

It is concave on the proximal end and convex on the distal end, where there are three facets for articulation with the cuneiforms.

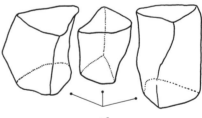

cuneiforms

The **cuneiforms** are three small wedge-shaped bones with the sharp edges directed inferiorly. They articulate proximally with navicular and distally with metatarsals I-III. Together with cuboid and the metatarsals, they constitute the transverse arch of the foot.

The anterior tarsals, and their many gliding joints, allow a reasonable degree of flexibility, though less than that of the corresponding wrist bones.

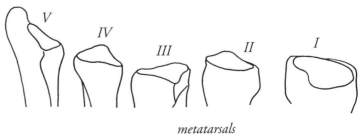

V   *IV*   *III*   *II*   *I*

metatarsals

LATERAL ⟵·········⟶ MEDIAL

Transverse tarsal joint (Chopart's joint)

This joint is located between the posterior and medial foot.
It comprises the articulation of the calcaneous
and cuboid bones laterally,
and the articulation of the talus
and the navicular bones medially.

talus

The medial part is
higher and involves
a convex surface of the
talar head fitting against
a concave surface
of navicular.

The lateral part is
lower and involves
an S-shaped surface
of calcaneus (concave
medially, convex laterally)
fitting against a corresponding
surface of cuboid.

calcaneus

cuboid

navicular

This shows the tarsal region from the front.
The two posterior bones are shown in anatomical position.
The two anterior bones, cuboid and navicular,
are tilted 90° to the back to show their posterior surface.

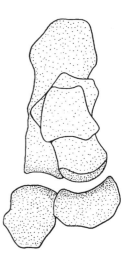

Seen from above,
the transverse tarsal joint
is curved rather than straight.

Midtarsal joint mobility

The basic movements here are inversion and eversion.
The dominant movement in this joint is **abduction-adduction**.

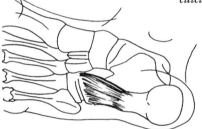

Capsules and ligaments of transverse tarsal joint

Superiorly, the joint is reinforced by the **talonavicular** and **calcaneocuboidal ligaments**. A medial capsule unites with the capsule of the anterior subtalar joint (see p. 269). A lateral capsule unites the calcaneous with the cuboid bone.

These capsules are reinforced by many ligaments.

Laterally, the **bifurcate ligament** runs from calcaneus and spreads out vertically on navicular and horizontally on cuboid (the surfaces of these two bones are roughly perpendicular at this junction).

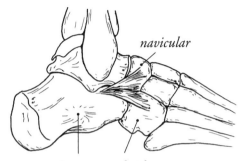

navicular

calcaneus   cuboid

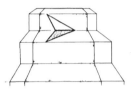

Inferiorly, the transverse tarsal joint is reinforced by three ligaments. The two layers of **plantar calcaneocuboid (short plantar) ligament** run from the calcaneus to the proximal cuboid, and to the base of the metatarsals.

The **long plantar ligament** (located superficial to the short plantar ligament) runs from calcaneus to a ridge on cuboid, passes over the groove (thereby forming a groove for the peroneus longus tendon), and continues on to attach to the bases of metatarsals II-V. The long plantar ligament is quite strong and helps support the arches of the foot.

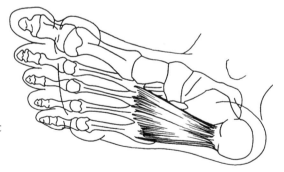

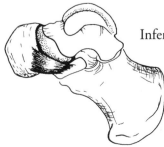

Inferomedially, the **plantar calcaneonavicular ligament** runs from the sustentaculum tall to navicular, and helps support the talar head.

Metatarsals and phalanges

These are five parallel "spokes" made of small bones, which flare out anteriorly. Every spoke consists of a **metatarsal**, as well as **phalanges**, which form the skeleton of the toe.

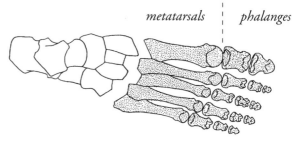

metatarsals　*phalanges*

base　　　*body*　　　*head*

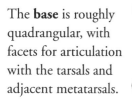

Each **metatarsal** consists of a proximal base, a body, and a distal head.

The **base** is roughly quadrangular, with facets for articulation with the tarsals and adjacent metatarsals.

The **head** is convexly rounded, with a cartilaginous surface for articulation with the proximal phalanx, and a tiny tubercle on each side.

The **body** is triangular in cross section, like most long bones.

The proximal **phalanx** of each toe has a concavely rounded base for articulation with the metatarsal, and a pulley-shaped head.

The base of the middle phalanx is concave but with a median crest to match the shape of the head of the proximal phalanx.

The head of the distal phalanx is flared to support the toenail superiorly, and has an inferior tubercle to support the fleshy part of the toe.

proximal phalanx

middle phalanx

distal phalanx

Tarsometatarsal joints

This group consists of the joints formed between the tarsal and metatarsal bones. This "articular line" joins the distal cuneiform and cuboid bones with the proximal bases of the metatarsal bones.

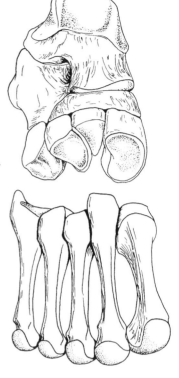

Collectively, the junction is irregular, not straight.

The articular surfaces make small gliding movements between the bones possible. As a group, the joints have some mobility, but it is quite limited.

Plantar flexion and dorsiflexion are the possible movements at these joints. ROM is in the increasing order II, III, I, IV, V.

The second spoke has very little mobility. It represents the axis of movement for pronation and supination of the foot.

These joints are supported by capsules, each of which is connected to neighboring capsules.

The capsules in turn are reinforced by many small ligaments, dorsal (shown at left) and plantar (not shown), which serve to connect the bones to one another.

Metatarsophalangeal joints

At each metatarsophalangeal joint the head of the metatarsal articulates with the base of the proximal phalanx at each of the five "spokes" of the forefoot. The articular surfaces have the shape of a small condyle, which allows movement in all three planes, as described on pages 8-10.

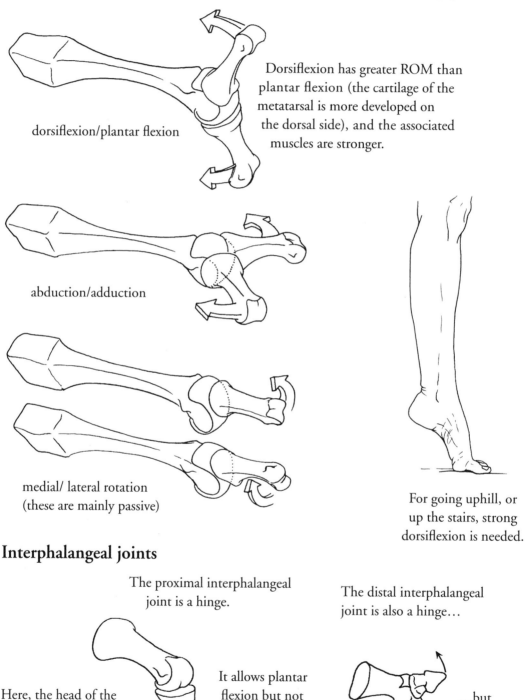

dorsiflexion/plantar flexion

Dorsiflexion has greater ROM than plantar flexion (the cartilage of the metatarsal is more developed on the dorsal side), and the associated muscles are stronger.

abduction/adduction

medial/ lateral rotation
(these are mainly passive)

For going uphill, or up the stairs, strong dorsiflexion is needed.

Interphalangeal joints

The proximal interphalangeal joint is a hinge.

The distal interphalangeal joint is also a hinge...

Here, the head of the proximal phalanx articulates with the base of the middle phalanx.

It allows plantar flexion but not dorsiflexion.

...but allows both plantar and dorsiflexion.

Particularities of 1st and 5th "spokes"

In the first spoke (metatarsal I), which includes the big toe, the metatarsal and phalanges are larger than in toes II-V, and there are two phalanges instead of three.

The big toe plays an important role in walking or running, especially in the digitigrade phase (i.e., when the toes are in contact with the ground).

The disproportionate size of metatarsal I can lead to instability or medial pain when on tiptoe or during prolonged walks.

Two small sesamoid bones are located in the plantar cartilage on the head of metatarsal I. They act as shock-absorbers during weightbearing.

In the fifth spoke, there is an externally-palpable tubercle (for muscle attachment) on the lateral base of metatarsal V.

Joint capsules and ligaments

Ligaments of the metatarsophalangeal and interphalangeal joints have the same general plan. The joints are held by a capsule, which is attached to the neighboring structures and reinforced by ligaments:

- two collateral ligaments, inserting on proximal tubercles of the distal bone

- a plantar "glenoid" ligament, which folds onto itself during plantar flexion

- a fan-shaped "deltoid" ligament, running from the tubercle to the glenoid ligament.

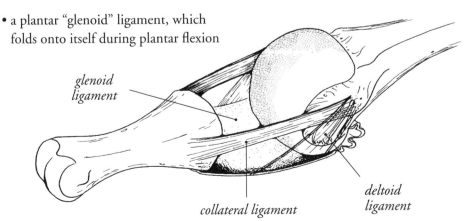

glenoid ligament

collateral ligament

deltoid ligament

Ankle and foot muscles with their many bony attachments

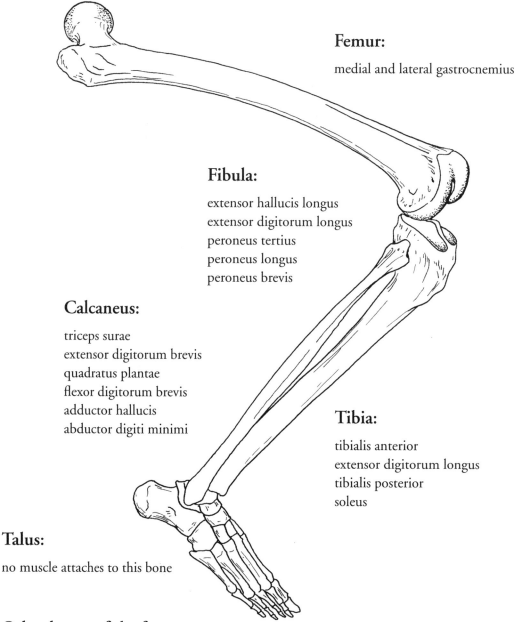

Femur:

medial and lateral gastrocnemius

Fibula:

extensor hallucis longus
extensor digitorum longus
peroneus tertius
peroneus longus
peroneus brevis

Calcaneus:

triceps surae
extensor digitorum brevis
quadratus plantae
flexor digitorum brevis
adductor hallucis
abductor digiti minimi

Tibia:

tibialis anterior
extensor digitorum longus
tibialis posterior
soleus

Talus:

no muscle attaches to this bone

Other bones of the foot:

All extrinsic muscles of the foot (except the triceps surae muscles)
and all the intrinsic muscles of the foot.

The **extrinsic muscles** of the foot originate on the femur, tibia, or fibula, and insert on the
bones of the foot via long tendons. They are all polyarticular, acting on the ankle and foot
(or on the knee, in the case of the gastrocnemius muscle). As they course down the leg, their
tendons run anterior or posterior to the ankle. The **intrinsic muscles** are short muscles which
run between nearby foot bones. They are mostly on the plantar side of the bones,
and comprise the fleshy mass of the sole.

Intrinsic muscles of the foot

Extensor digitorum brevis
is the only dorsal intrinsic muscle.

It arises from the anterosuperolateral calcaneus and divides into four bodies.

The medial tendon inserts on proximal phalanx I, while the other three merge laterally with the tendons of extensor digitorum longus, inserting on toes II-IV.

Actions: dorsiflexion of toes I-IV, especially at the level of the proximal phalanx; reinforces the action of extensor digitorum longus

Innervation: deep peroneal nerve (S1-S2)

Middle group

On the sole of the foot, the intrinsic muscles can be subdivided into three groups: middle, medial, and lateral. These first two pages describe the middle group. Collectively, these muscles occupy several layers, but each of the following illustrations shows each muscle layer separately.

The **interossei** form the fourth (deepest) layer, and occupy the spaces between the metatarsals, where they originate.

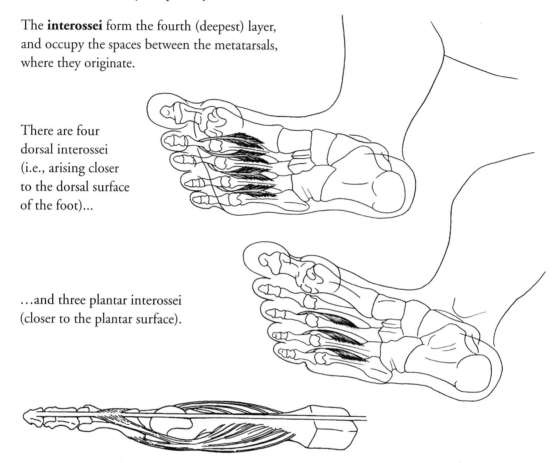

There are four dorsal interossei (i.e., arising closer to the dorsal surface of the foot)...

...and three plantar interossei (closer to the plantar surface).

They insert at the base of the proximal phalanges (plantar side) and at the extensor digitorum longus tendon (dorsal side, as shown in simplified drawing above).

Action: principally, plantar flexion of proximal phalanges...

...which is important in the propulsion phase of walking

Innervation: lateral plantar nerve (S1–S2)

By contracting on one side or the other, the interossei also help spread the toes apart or pull them back together. Their origins (proximal attachments) help prevent the metatarsals from spreading apart, and maintain the transverse arch of the foot.

The interossei are covered by the tendons of flexor digitorum longus. At the hindfoot, a muscle attaches to these tendons:

quadratus plantae

lumbricals

Quadratus plantae (also called flexor digitorum accessorius) arises from the body of calcaneus, and inserts on the posterolateral border of the flexor digitorum longus tendon near its division into four parts. It belongs to the second layer.

The **lumbricals** are four small muscles (in the second layer) running from the flexor digitorum longus tendons to the dorsal parts of the extensor digitorum longus tendons.

Action: minimal; they mostly "fine tune" the actions of other toes muscles

Innervation: medial and lateral plantar nerves (L5-S2)

Actions: redirects the pull of the flexor digitorum longus tendons to be more in line with the axes of the toes

Innervation: lateral plantar nerve (S1-S2)

Flexor digitorum brevis is part of the first (most superficial) layer.

It arises from the posteroinferior tuberosity of calcaneus, splits into four parts, and inserts laterally on middle phalanges II-V. Each tendon is "perforated" to allow passage of the flexor digitorum longus tendons to the distal phalanges.

Action: plantar flexion of middle and proximal phalanges of toes II-V; often responsible for the condition called "clawfoot," particularly when action of the interossei is weak

Innervation: medial plantar nerve (L5-S1)

Medial group

There are three muscles which insert on the proximal phalanx of the big toe and, in passing, on the sesamoid bone.

Flexor hallucis brevis (third layer) originates from cuboid and the two lateral cuneiforms and inserts via two tendons on either side of proximal phalanx I.

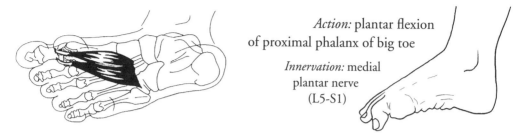

Action: plantar flexion of proximal phalanx of big toe

Innervation: medial plantar nerve (L5-S1)

Adductor hallucis has two layers. The oblique adductor arises on the cuboid, and the transverse adductor on the metatarsaophalangeal joints III-V. Both layers merge in a tendon, which inserts on the lateral base of proximal phalanx I.

This is the muscle that is primarily responsible for "hallux valgus," a malformation in which metatarsal I is permanently adducted and prosimal phalanx I abducted, such that the big toe overlaps the second toe.

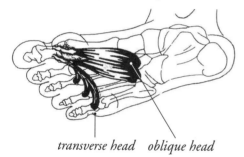

transverse head   oblique head

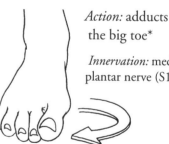

Action: adducts the big toe*

Innervation: medial plantar nerve (S1-S2)

Abductor hallucis is the most superficial of the three muscles. It originates at the medial tuberosity on the medial surface of the calcaneous and inserts on the medial base of the proximal phalanx I.

This muscle actively supports the medial arch and assists in keeping the big toe properly aligned by opposing "hallux valgus."

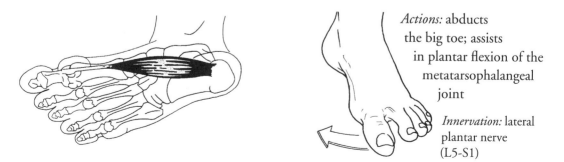

Actions: abducts the big toe; assists in plantar flexion of the metatarsophalangeal joint

Innervation: lateral plantar nerve (L5-S1)

Note: When we speak of adduction or abduction of toes, the reference is the axis of the 2nd toe, not the median plane of the body.

Lateral group

There are three small muscles on the lateral side of the foot.

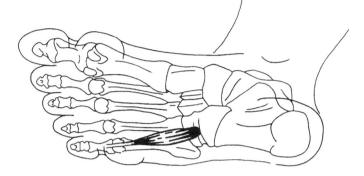

Flexor digiti minimi brevis (third layer) arises from the base of metatarsal V and inserts on the base of proximal phalanx V.

Action: plantar flexion of little toe

Innervation: lateral plantar nerve (S1-S2)

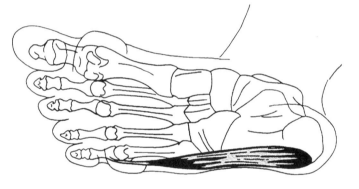

Abductor digiti minimi (first layer) originates from the posteroinferior calcaneus and inserts laterally on the base of proximal phalanx V.

Actions: abduction and plantar flexion of little toe; supports lateral arch

Innervation: lateral plantar nerve (S1-S2)

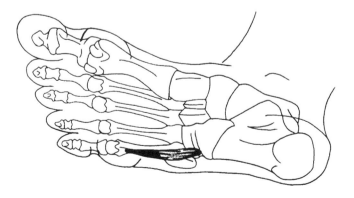

Opponens digiti minimi originates from cuboid and inserts on the lateral side of metatarsal V.

Actions: orients metatarsal V toward the other metatarsals and resists the spreading of the anterior portion of the foot

Innervation: lateral plantar nerve (S1-S2)

Extrinsic anterior muscles

On the anterior surface of the leg are three long muscles whose tendons run anterior to the ankle, where they are held in place by the **extensor retinaculum**.

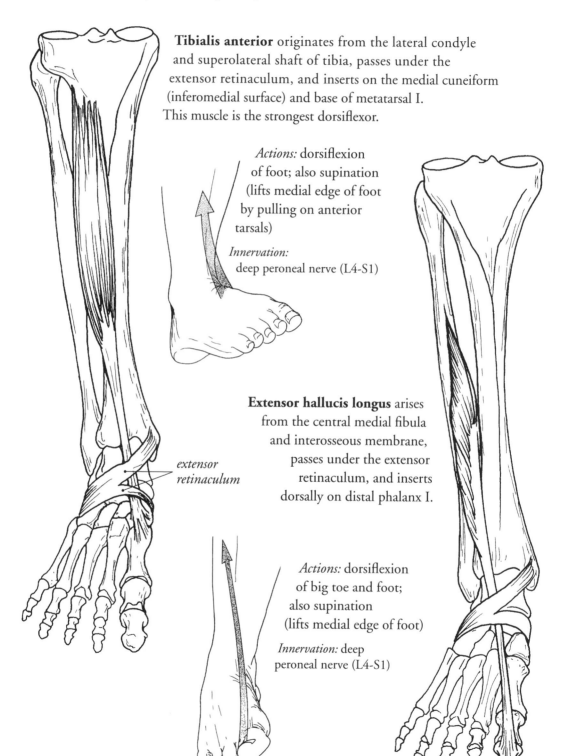

Tibialis anterior originates from the lateral condyle and superolateral shaft of tibia, passes under the extensor retinaculum, and inserts on the medial cuneiform (inferomedial surface) and base of metatarsal I. This muscle is the strongest dorsiflexor.

Actions: dorsiflexion of foot; also supination (lifts medial edge of foot by pulling on anterior tarsals)

Innervation: deep peroneal nerve (L4-S1)

extensor retinaculum

Extensor hallucis longus arises from the central medial fibula and interosseous membrane, passes under the extensor retinaculum, and inserts dorsally on distal phalanx I.

Actions: dorsiflexion of big toe and foot; also supination (lifts medial edge of foot)

Innervation: deep peroneal nerve (L4-S1)

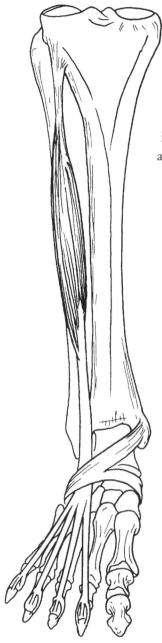

Extensor digitorum longus originates from the lateral tibial condyle, most of the anterior fibular shaft, and interosseous membrane. Its tendon passes under the extensor retinaculum, splits into four parts, and inserts on toes II-V. Each of the four tendons further splits into two slips attaching to the sides of the middle phalanx, and a central slip attaching to the base of the distal phalanx.

Action: dorsiflexion of toes II-V, foot, and ankle; it mainly acts on the proximal phalanx and is one of the muscles responsible for the "clawing" action of the toes

Innervation: deep peroneal nerve (L4-S1)

Short muscles of the foot that insert on the tendon of extensor digitorum longus, complementing its action:

- extensor digitorum brevis (p. 281)
- interossei (p. 282)

Peroneus tertius is an insignificant muscle, absent in some individuals. It arises from the anteroinferior fibula and inserts on metatarsal V.

Actions: dorsiflexion and eversion of foot

Innervation: deep peroneal nerve (L5-S1)

Extrinsic lateral muscles

There are two muscles on the lateral side of the leg that attach to the fibula: the **peroneus muscles**.

Peroneus longus arises from the head and superolateral shaft of fibula.

Its tendon follows a complicated path behind the lateral malleolus, under the peroneal retinaculum, *inferior* to the peroneal tubercle of calcaneus, along the groove of the plantar (inferior) cuboid, and finally inserts inferiorly on the base of the medial cuneiform and base of metatarsal I.

Peroneus brevis arises from the inferolateral fibular shaft, where it is covered by peroneus longus.

Its tendon passes behind the lateral malleolus, under the peroneal retinaculum, *superior* to the peroneal tubercle, and inserts on the lateral tubercle of metatarsal V.

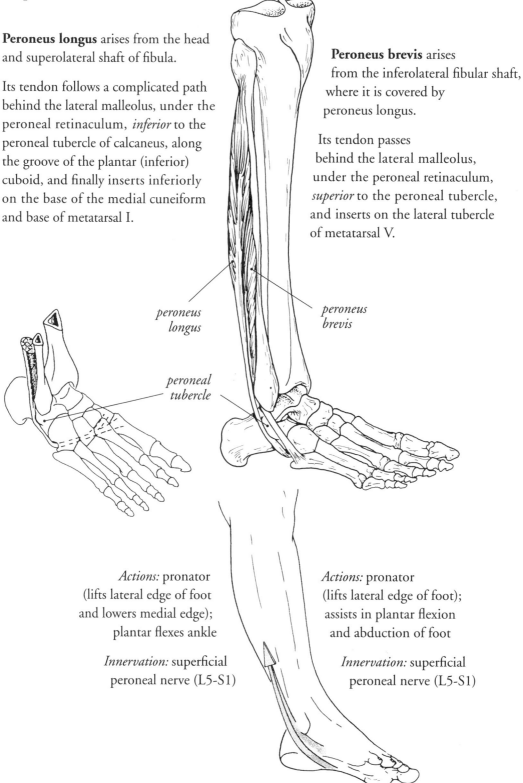

*peroneus
longus*

*peroneus
brevis*

*peroneal
tubercle*

Actions: pronator
(lifts lateral edge of foot
and lowers medial edge);
plantar flexes ankle

Innervation: superficial
peroneal nerve (L5-S1)

Actions: pronator
(lifts lateral edge of foot);
assists in plantar flexion
and abduction of foot

Innervation: superficial
peroneal nerve (L5-S1)

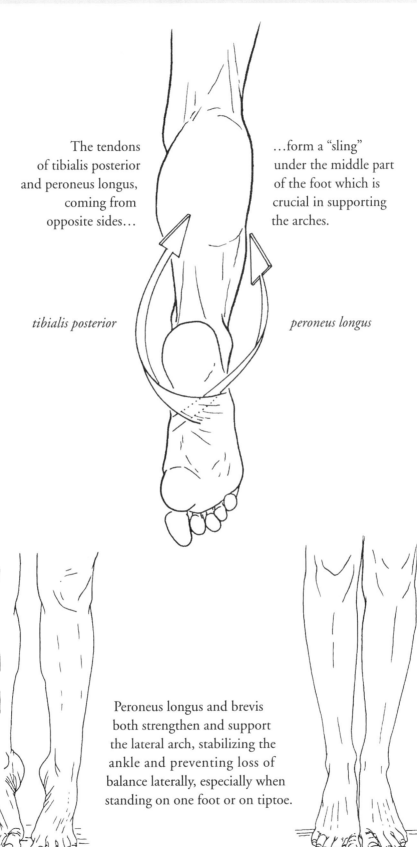

The tendons of tibialis posterior and peroneus longus, coming from opposite sides…

…form a "sling" under the middle part of the foot which is crucial in supporting the arches.

tibialis posterior

peroneus longus

Peroneus longus and brevis both strengthen and support the lateral arch, stabilizing the ankle and preventing loss of balance laterally, especially when standing on one foot or on tiptoe.

Extrinsic posterior muscles

The posterior group of leg muscles is the most important muscle group. It has two layers, the deeper one consisting of three muscles located next to each other on the posterior surfaces of the tibia and fibula.

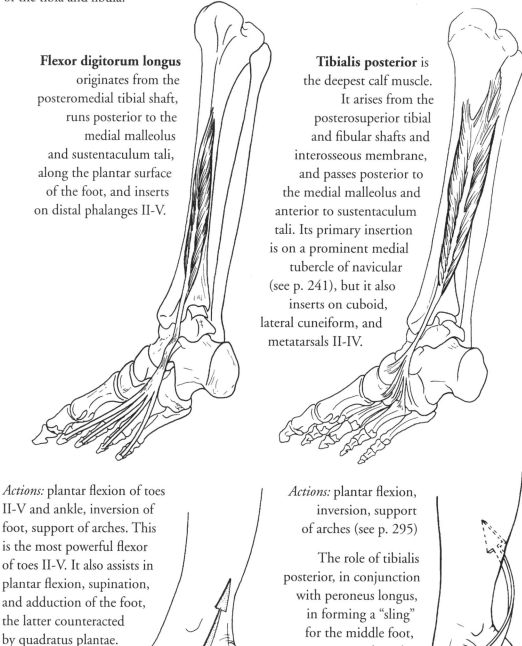

Flexor digitorum longus originates from the posteromedial tibial shaft, runs posterior to the medial malleolus and sustentaculum tali, along the plantar surface of the foot, and inserts on distal phalanges II-V.

Tibialis posterior is the deepest calf muscle. It arises from the posterosuperior tibial and fibular shafts and interosseous membrane, and passes posterior to the medial malleolus and anterior to sustentaculum tali. Its primary insertion is on a prominent medial tubercle of navicular (see p. 241), but it also inserts on cuboid, lateral cuneiform, and metatarsals II-IV.

Actions: plantar flexion of toes II-V and ankle, inversion of foot, support of arches. This is the most powerful flexor of toes II-V. It also assists in plantar flexion, supination, and adduction of the foot, the latter counteracted by quadratus plantae.

Innervation: tibial nerve (S1-S3)

Actions: plantar flexion, inversion, support of arches (see p. 295)

The role of tibialis posterior, in conjunction with peroneus longus, in forming a "sling" for the middle foot, was noted on the preceding page.

Innervation: tibial nerve (L4-L5)

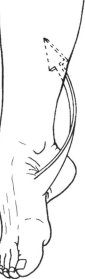

Flexor hallucis longus arises from the posteroinferior fibula and interosseous membrane, runs posterior to the medial malleolus, along a groove on the posterior talus, behind sustentaculum tali, along the medial plantar surface of the foot, and inserts on distal phalanx I.

This muscle is important in the propulsion phase of walking.

Actions: plantar flexion of big toe and ankle, inversion, support of medial arch; assists in plantar flexion and adduction of foot

Innervation: tibial nerve (S1-S3)

It also has a very important role in stabilizing the foot when standing on tiptoe, since the pushing action of the big toe offsets the anterior loss of balance. It also helps stabilize the ankle (see p. 295).

The superficial group of posterior calf muscles is known collectively as the **triceps surae**.

This is the strongest muscle group of the leg.
It consists of three muscular heads,
all of which terminate on the **Achilles tendon**,
which then attaches to the posterior surface of the calcaneus.

Soleus is the
deepest head
of triceps surae.
It arises from the
posterosuperior
tibia and fibula.

Soleus is covered by the
two superficial heads of
gastrocnemius, which
originate at the distal
posterior femur from a
tendon attaching to the
back of each condyle.

It crosses two joints:
the ankle and the
subtalar joint.

Gastrocnemius shapes the
outline of the posterior calf.
It crosses three articulations:
the ankle, subtalar joint, and knee.

Innervation: common
tibial nerve (L5/S2)

Innervation: common
tibial nerve (S1-S2)

Actions of triceps surae: Together,
the three muscles pull the
calcaneus into plantar flexion
under the talus, with a
tendency to inversion*…

…and indirectly pull the talus into
plantar flexion. This second action
is actually more important than
the first because it gives the joint
more mobility.

*Why inversion? It is linked to the articular surfaces of the subtalar joint.
Plantar flexion corresponds to adduction and supination (see p. 271).

Since gastrocnemius crosses the knee, the position of the knee affects its efficiency as a plantar flexor of the ankle.

For standing on tiptoe, plantar flexion of the ankle by the triceps surae is necessary, but not sufficient by itself.

When the knee is very flexed, the gastrocnemius is slack and therefore inefficient as an ankle flexor.

When the knee is extended or only slightly flexed (the position taken by the "propelling leg" at the start of a race), the gastrocnemius is more taut and more efficient as an ankle flexor.

Dorsiflexion at the ankle stretches the soleus.

Interestingly, when the knee is flexed and the leg and foot are bearing the weight of the body…

…the gastrocnemius and hamstrings combine their forces to become *extensors* of the knee, i.e., returning it to anatomical position.

When the foot is not bearing the body's weight, these muscles act as *flexors* of the knee.

To stretch the gastrocnemius, we must add extension of the knee.

Actions of extrinsic foot muscles on the ankle

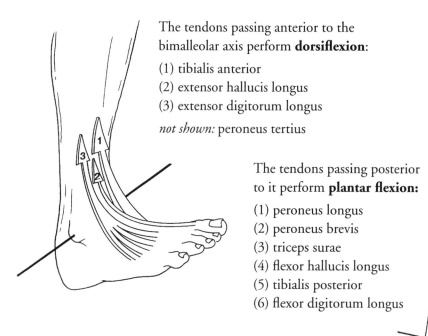

The tendons passing anterior to the bimalleolar axis perform **dorsiflexion**:

(1) tibialis anterior
(2) extensor hallucis longus
(3) extensor digitorum longus

not shown: peroneus tertius

The tendons passing posterior to it perform **plantar flexion**:

(1) peroneus longus
(2) peroneus brevis
(3) triceps surae
(4) flexor hallucis longus
(5) tibialis posterior
(6) flexor digitorum longus

The tendons passing medial to the longitudinal axis of the foot (axis through 2nd toe) perform **inversion/adduction**:

(1) extensor hallucis longus
(2) tibialis anterior
(3) tibialis posterior
(4) flexor digitorum longus
(5) flexor hallucis longus

We could add triceps surae, whose action adds an inversion (see p. 292).

The tendons passing lateral to the longitudinal axis of the foot perform **eversion/abduction**:

(1) peroneus longus and brevis
(2) peroneus tertius
(3) extensor digitorum longus (lateral part)

Notice that opposing actions are not "balanced." Plantar flexion is dominant over dorsiflexion, and inversion/adduction is dominant over eversion/abduction.

Stability of ankle joint

As explained on page 264, the talus fits more snugly into the "pincer" (formed by the distal tibia and fibula) during dorsiflexion than plantar flexion.

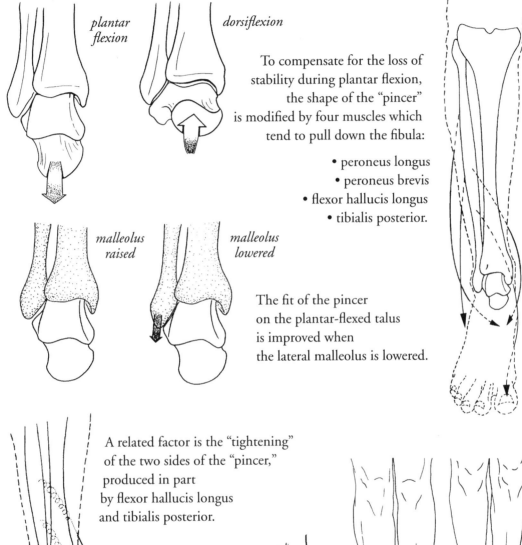

plantar flexion

dorsiflexion

malleolus raised

malleolus lowered

To compensate for the loss of stability during plantar flexion, the shape of the "pincer" is modified by four muscles which tend to pull down the fibula:

- peroneus longus
- peroneus brevis
- flexor hallucis longus
- tibialis posterior.

The fit of the pincer on the plantar-flexed talus is improved when the lateral malleolus is lowered.

A related factor is the "tightening" of the two sides of the "pincer," produced in part by flexor hallucis longus and tibialis posterior.

Also, when the fibula is lowered, the distal tibiofibular ligaments come under tension, which automatically tightens the "pincer."

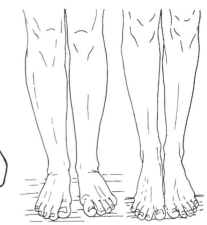

These effects on the "pincer" by muscles and ligaments stabilize the ankle during plantar flexion, e.g., standing on tiptoe.

Arches of the foot

The foot has three arches (which could also be called "trusses"*).
They rest on three points of support. The arches therefore constitute a flexible slat,
which serves as a shock absorber and adapts the foot's shape to the ground.

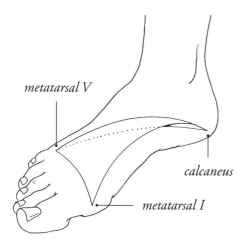

When standing, the weight of the body
is essentially distributed among three points:

- The posteroinferior tuberosity of calcaneus
  receives most of the weight.

- The secondary weight-bearing point
  is the head of metatarsal I,
  a relatively massive bone.

- The third point, the head of metatarsal V,
  supports the least weight.

The **medial arch** is formed primarily by five bones (calcaneus,
talus, navicular, medial cuneiform, metatarsal I), four ligaments
(talocalcaneal, calcaneonavicular, and small ligaments joining
the cuneiform to navicular and metatarsal), and four
muscles (abductor hallucis, tibialis posterior,
peroneus longus, flexor hallucis longus).

Flexor hallucis longus has three functions
with respect to the medial arch:

1. stretches the arch like the string of a bow
2. supports calcaneus by passing under the sustentaculum tali
3. supports talus by passing along its posterior groove.

*In architecture, a supporting
structure in the shape of
a triangle (a truss) has
the following stress
distribution:

The weight on the top part causes compression stress (on the top) and stretching stress (on the
bottom). Due to the relative elasticity of the lower part, a very heavy weight can be supported.

The **lateral arch** is not as high as the medial one. Although it can be plainly seen on the skeleton, it is not obvious on the whole foot because the space under the arch is occupied by muscles.

The lateral arch is formed primarily by three bones (calcaneus, cuboid, metatarsal V), three ligaments (short plantar [calcaneocuboid] ligament, long plantar ligament, plantar aponeurosis), and two muscles (peroneus brevis, peroneus longus).

Peroneus longus has two functions here:
- supports calcaneus by passing under the peroneal tubercle
- supports cuboid.

The **transverse arch** is most visible around the middle of the metatarsals. It is represented in this drawing as straps.

As you would expect, it is higher on the medial (navicular) side than the lateral (cuboid) side.

Its muscular support comes primarily from adductor hallucis (transverse head), peroneus longus, tibialis posterior, and the interossei.

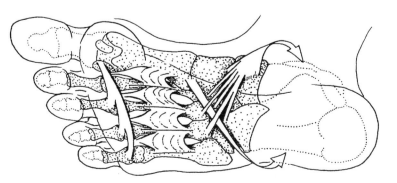

Muscle actions of ankle and foot during walking

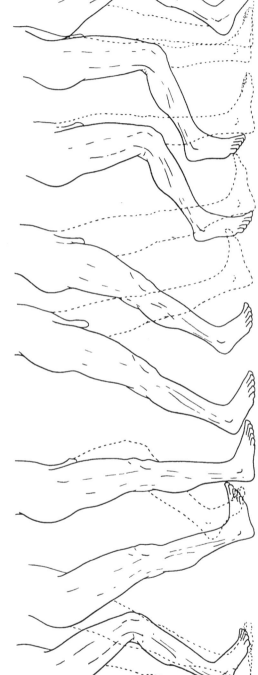

foot hits the ground heel-first, then rolls forward

body's weight on foot, foot in full contact with ground

heel leaves ground; propulsion by distal foot

toes leave ground (big toe last)

entire foot is off the ground; foot moves rapidly forward

contraction of dorsiflexors during rolling movement

contraction of muscles supporting the three arches

contraction of triceps surae and other plantar flexors; contraction of intrinsic plantar muscles

contraction of flexor digitorum longus, then flexor hallucis longus

brief moment of muscular relaxation while foot is off ground; contraction of dorsiflexors to prevent toes from touching ground during forward movement of foot

Index

L

N

O

S